Algorithmic Cultures

This book provides in-depth and wide-ranging analyses of the emergence, and subsequent ubiquity, of algorithms in diverse realms of social life. The plurality of *Algorithmic Cultures* emphasizes: (1) algorithms' increasing importance in the formation of new epistemic and organizational paradigms; and (2) the multi-faceted analyses of algorithms across an increasing number of research fields. The authors in this volume address the complex interrelations between social groups and algorithms in the construction of meaning and social interaction. The contributors highlight the performative dimensions of algorithms by exposing the dynamic processes through which algorithms—themselves the product of a specific approach to the world—frame reality, while at the same time organizing how people think about society. With contributions from leading experts from Media Studies, Social Studies of Science and Technology, Cultural and Media Sociology from Canada, France, Germany, UK and the USA, this volume presents cutting-edge empirical and conceptual research that includes case studies on social media platforms, gaming, financial trading and mobile security infrastructures.

Robert Seyfert is a Postdoctoral Fellow at the Cluster of Excellence "Cultural Foundations of Social Integration" at Universität Konstanz, Germany, and Visiting Full Professor of Comparative Cultural Sociology at Europa-Universität Viadrina Frankfurt (Oder), Germany. He recently published in *Theory, Culture & Society* and *European Journal of Social Theory*.

Jonathan Roberge is Assistant Professor of Cultural and Urban Sociology at the Institut National de la Recherche Scientifique, Quebec; he holds the Canada Research Chair in Digital Culture, in addition to being a Faculty Fellow at the Center for Cultural Sociology at Yale University.

Routledge Advances in Sociology

Algorithmic Cultures

Essays on meaning, performance and new technologies

**Edited by Robert Seyfert and
Jonathan Roberge**

Routledge
Taylor & Francis Group

LONDON AND NEW YORK

First published 2016 by Routledge

2 Park Square, Milton Park, Abingdon, Oxfordshire OX14 4RN
711 Third Avenue, New York, NY 10017

Routledge is an imprint of the Taylor & Francis Group, an informa business

First issued in paperback 2018

British Library Cataloguing-in-Publication Data
A catalogue record for this book is available from the British Library

Library of Congress Cataloging-in-Publication Data
A catalog record for this book has been requested

ISBN: 978-1-138-99842-1 (hbk)
ISBN: 978-1-138-35189-9 (pbk)

Typeset in Times New Roman
by Wearset Ltd, Boldon, Tyne and Wear

Every effort has been made to source permissions for the figures in
the book.

Contents

Figures

Contributors

Jean-Samuel Beuscart is a Sociologist at Orange Labs and Associate Professor at the University of Marne-la-Vallée (LATTS). He is currently working on the framing of Internet markets as well as the implications of online visibility. He published *Promouvoir les oeuvres culturelles* (Paris: La Documentation Française, 2012), with Kevin Mellet. With Dominique Cardon, he now leads the project "Algopol," which receives substantial support from the Agence Nationale de la Recherche in France.

Dominique Cardon is a Sociologist in the Laboratory of Uses of France Telecom R&D and Associate Professor at the University of Marne-la-Vallée (LATTS). He is studying transformations of public space and the uses of new technologies. He has published different articles on the place of new technologies in the no-global movement, alternative media and on the process of bottom-up innovations in the digital world. He published *La démocratie Internet* (Paris: Seuil/République des idées, 2010) and, with Fabien Granjon, *Médiactivistes* (Paris: Presses de Science Po, 2010).

Tarleton Gillespie is an Associate Professor at the Department of Communication at Cornell University and is currently a visitor with Microsoft Research New England. He is the author of *Wired Shut: Copyright and the Shape of Digital Culture* (Cambridge, MA: MIT Press, 2007) and the co-editor (with Pablo Boczkowski and Kirsten Foot) of *Media Technologies: Essays on Communication, Materiality, and Society* (Cambridge, MA: MIT Press, 2014). He is also a co-founder (with Hector Postigo) of the NSF-sponsored scholarly collective Culture Digitally (culturedigitally.org). He is currently finishing a book on the implications of the content policies of online platforms for Yale University Press, and has written on the relevance of algorithms for the changing contours of public discourse.

Lucas D. Introna is Professor of Technology, Organization and Ethics at the Centre for the Study of Technology and Organization, Lancaster University. His primary research interest is the social study of technology. In particular, he is concerned with theorizing social/technical entanglements, especially with regard to ethics and politics. He has published on a variety of topics, such as sociomateriality, performativity, phenomenology of technology, information and power, privacy, surveillance, technology ethics and virtuality. He is a co-editor of *Ethics and Information Technology* and has acted as associate editor for a variety of leading journals.

Joseph Klett is a Visiting Assistant Professor in the Department of Sociology at the University of California, Santa Cruz (PhD Yale), and a regular participant in the American digitalSTS initiative. He has recently written two articles, "The Meaning of Indeterminacy" about the social practices which lend meaning to Noise Music (*Cultural Sociology*, 2014), and "Sound on Sound," about the ethnographic study of sonic interactions (*Sociological Theory*, 2014).

Oliver Leistert is a media and technologies scholar at Leuphana Universität Lüneburg, Germany. Previously he was a Postdoctoral Researcher at the "Automatisms" research group, University of Paderborn. He is a collaborator at the ESRC project "Digital Citizenship and Surveillance Society" at the University of Cardiff. He has studied philosophy, computer science and literature. His doctoral thesis in media studies, "From Protest to Surveillance: The Political Rationality of Mobile Media" won the Surveillance & Society Book Award in 2014. Other recent publications include (co-edited with Lina Dencik) *Critical Perspectives on Social Media and Protest: Between Control and Emancipation* (London: Rowman & Littlefield, 2015).

Kevin Mellet is a Researcher at the Social Sciences Department of Orange Labs and Associate Researcher at the Centre de Sociologie de l'Innovation (Mines ParisTech). Originally trained as an economist, he has developed expertise in economic sociology and science and technology studies. His research explores the construction of digital markets. Current areas of interest include marketing and advertising practices, participatory valuation devices, business models and market intermediation. He is the co-author (with Jean-Samuel Beuscart) of a book on advertising strategies within cultural industries (*Promouvoir les œuvres culturelles*, Paris: La Documentation Française, 2012).

Shintaro Miyazaki is a Senior Researcher and Lecturer at the University of Applied Sciences and Arts Northwestern Switzerland, Academy of Art and Design, Institute of Experimental Design and Media Cultures. He is writing at the intersection of media history, design theory and the history of science and technology. Previously, he was a Resident Fellow at the Akademie Schloss Solitude in Stuttgart (2011–2012) and Art/Science Resident at the National University of Singapore (September 2012). He not only works as a scholar, but also actively engages in practices between experimental design and artistic research.

Valentin Rauer works as a Senior Research Fellow at the Cluster of Excellence "The Formation of Normative Orders" at Frankfurt University. He is interested in social and cultural processes that transform, transmit and translate the past (collective memories and identities), and the future (security cultures and risks). Current publications include "The Visualization of Uncertainty," in *Iconic Power: Materiality and Meaning in Social Life*, ed. Alexander, Jeffrey C. *et al.* (New York: Palgrave Macmillan, 2012) and "Von der Schuldkultur zur Sicherheitskultur: Eine begriffsgeschichtliche Analyse 1986–2010," *Sicherheit & Frieden* (February 2011).

Jonathan Roberge is Assistant Professor of Cultural and Urban Sociology at the Institut National de la Recherche Scientifique, Quebec; he holds the Canada Research Chair in Digital Culture, in addition to being a Faculty Fellow at the Center for Cultural Sociology at Yale University.

Robert Seyfert is a Postdoctoral Fellow at the Cluster of Excellence "Cultural Foundations of Social Integration" at Universität Konstanz, Germany, and Visiting Full Professor of Comparative Cultural Sociology at Europa-Universität Viadrina Frankfurt (Oder), Germany. He recently published in *Theory, Culture & Society* and *European Journal of Social Theory*.

Acknowledgments

This volume evolved out of work initially presented at the Algorithmic Cultures Conference at University of Konstanz in Germany, June 23–25, 2014. This volume and the conference were made possible with the generous support of the Canada Research Chairs Program and the "Cultural Foundations of Social Integration" Centre of Excellence at the University of Konstanz, established in the framework of the German Federal and State Initiative for Excellence.

1   What are algorithmic cultures?

Jonathan Roberge and Robert Seyfert

The current, widespread dissemination of algorithms represents a double challenge for both our society and the social sciences tasked with studying and making sense of them. Algorithms have expanded and woven their logic into the very fabric of all social processes, interactions and experiences that increasingly hinge on computation to unfold; they now populate our everyday life, from the sorting of information in search engines and news feeds, to the prediction of personal preferences and desires for online retailers, to the encryption of personal information in credit cards, and the calculation of the shortest paths in our navigational devices. In fact, the list of things they can accomplish is rapidly growing, to the point where no area of human experience is untouched by them—whether the way we conduct war through ballistic missile algorithms and drones, or the manner in which we navigate our love lives via dating apps, or the way we choose how to dress by looking at weather forecasts. Algorithms make all of this possible in a way that initially appears disarmingly simple. One way to approach algorithms is through Kowalski's now classic definition: "Algorithm = Logic + Control" (1979). Using both simple and complex sorting mechanisms at the same time, they combine high-level description, an embedded command structure, and mathematical formulae that can be written in various programming languages. A wide variety of problems can be broken down into a set of steps and then reassembled and executed or processed by different algorithms. Hence, it is their versatility that constitutes their core capability and power, which extends far beyond the mathematical and computer sciences. According to Scott Lash, for instance, "a society of ubiquitous media means a society in which power is increasingly in the algorithms" (2007, 71), an idea echoed by Galloway when he states that "the point of power today resides in networks, computers, algorithms, information and data" (2012, 92). Yet, it is imperative to remain cautious with such formulations, and their tendency to be too critical, too quickly. While it may capture important challenges that society faces with 'the rise of the algorithm,' it can also provide something of a teleological or deterministic "seductive drama," as Zietwitz has recently warned us (2016, 5). Algorithms can actually be considered less sovereign than mundane in this regard—that is, again, deeply rooted in the fabric of society. Rather than being omnipotent, they are oftentimes ambiguous and quite messy. What is crucial, then, is to bring into question how, and especially *why*, the apparent simplicity of algorithms is in fact inseparable from their complexity, in terms of their deployment and multiple, interrelated ramifications. These are epistemological as well as ontological interrogations, confronting not only the social sciences but society at large. As both a known unknown and an unknown known, the sorting mechanism that is the algorithm still needs some sorting out.

This introduction is certainly not the first to stress the inherent difficulty of shedding light on algorithms. Seaver, for instance, observes how they "are tricky objects to know" (2014, 2), while Sandvig insists on "the complexity of representing algorithms" (2015, 1; see also Introna 2016; Barocas *et al.* 2013). Conceptually perspicacious as they are, these arguments do not, however, foreclose the need to understand the extent of such invisibility and inscrutability. On the surface, it is often the 'black box' nature of the algorithms that is first evoked, namely that they are incredibly valuable patented trade secrets for companies such as Amazon, Google, Facebook, and the like. If they were revealed to non-insiders, they would *eo ipso* be ruined. Or at least so we are told by numerous technical, economic, legal, and political experts (Pascale 2015). This is where things noticeably start to get more serious and profound. There is not one box, but multiple boxes. The opacity of algorithms is more precisely expressed in different forms of opacity, all of which, in specific ways, are contingent on the *in-betweenness* of a plethora of actors, both human and non-human. While a few commentators have remarked upon the plural character of such opacity (Burrell 2016; Morris 2015), the fact remains that each and every algorithm can only exist in rich and dense, if not tense, environments.

This is the inherently messy, vivid, and dynamic nature of algorithms, which explains why they are ultimately so challenging to study. As Kitchin puts it, "creating an algorithm unfolds in context through processes such as trial and error, play, collaboration and negotiation" (2014, 10). The latter term is of particular interest here: "negotiation" refers to the very condition of possibility/difficulty of algorithms. On the most fundamental level, they are what one can call *anthropologically entrenched* in us, their creators and users. In other words, there is a "constitutive entanglement" where "it is not only us that make them, they also make us" (Introna and Hayes 2011, 108). Indeed, the problem with such mutual imbrication is that algorithms cannot be fully 'revealed,' but only unpacked to a certain extent. What is more, they always find themselves *temporally entrenched*, so to speak. They come to life with their own rhythm, or, to use Shintaro Miyazaki's description in this volume, "they need unfolding, and thus they embody time" (p. 129). Another metaphor that proves useful in this regard is Latour's idea of the cascade (1986, 15–16): algorithms follow a non-linear course, caught in constant changes, fluctuations, and deviations both large and small. Such changes may very well be hard to follow or may even be imperceptible from time to time. The most important point to make here is how practical and mundane they are. Again, they unfold in a state of incessant negotiation and in-betweenness; for all algorithms, as Seaver has noticed, there are "hundreds of hands reaching into them, tweaking and tuning, swapping out parts and experiencing with new arrangements" (2014, 10).

The multiple ways in which algorithms unfold today thus give new meaning to the familiar description, "the most profound technologies are those that disappear" (Weiser 1991, 95). But there is more. We would like to take this opportunity to argue that such concrete unfoldings also give a new yet complex meaning to what it is that algorithms actually *do*, i.e., the kind of agency and performativity they embody. Of course, there is now a substantial tradition of academics working within this broadly defined praxiological paradigm, including Lucas Introna (this volume, 2016, 2011), Adrian Mackenzie (2005), David Beer (2013), and Solon Barocas *et al.* (2013). Somewhat aligning ourselves with them, we invoke Andrew Goffey's persuasive insight that "algorithms do things, and their syntax embodies a command structure to enable this to happen" (2008, 17)—an insight almost as persuasive as Donald MacKenzie's description of the algorithm as "an engine, not a camera" (2006). Many things could be said about such a position, and it will be important to come back to them in due time. It suffices for the moment to say that the agency of algorithms is a far cry from the category of 'action,' if we understand by the latter something purposive and straightforward. On the contrary, the type of agency involved here can be best described as 'fractal,' that is, producing numerous outputs from multiple inputs (Introna 2016, 24). What counts as 'control' in the algorithmic sense is in fact relatively limited; there is so much more implied before, during, and after the operation of algorithms. For instance, to both the anthropological and temporal entrenchment discussed above, it appears necessary to add the concept of *self-entrenchment*, whereby one algorithm is intertwined with many others in extremely intricate networks. Non-human as much as human contributions are thus key here, and could rather easily result in mismatches, unpredictable results, or even dramatic failure—as will be seen later. It is as if algorithms themselves are constituted by the very possibility of 'being lost in translation,' not only in their relations to machines, code, or even some more discursive dimensions, but in terms of the entire practicality and performativity that defines them. For an algorithm is performative by definition, and to be performative is to be heterogeneous in all circumstances (Kitchin 2014, 14–15; Seaver 2014). To be able to carefully read such messy unfoldings constitutes a pressing challenge for the social sciences in general, and for cultural sociology in particular. What does it mean, indeed, if these unfoldings themselves become a particular object of investigation? How is it that we could or should adapt in turn, with what kind of precision, changes in focus, and so forth?

Now is an appropriate moment to assess the state of research on algorithms in the so-called 'soft sciences,' and to reflect on both its virtues and shortcomings. The fact is that the field of algorithmic research has arrived at a certain degree of maturity, even if it was not until very recently that it started to migrate to the humanities, social sciences, and cultural studies. Currently, there are several promising cross-currents that more or less co-exist, but that do not yet properly engage with one another. First, there are those authors developing almost stand-alone concepts: "the algorithmic turn" (Uricchio 2011), "algorithmic ideology" (Mager 2012), "algorithmic identity" (Cheney-Lippold 2011), "algorithmic life"

(Amoore and Piotukh 2016), and the like. There are also significant attempts toward a 'sociology of algorithms' that have emerged in the field of Science and Technologies Studies (STS) and the Social Studies of Finance (MacKenzie 2015; Wansleben 2012), as well as embryonic efforts to develop Critical Algorithm Studies (The Social Media Collective 2015). In addition, there have been several important conferences over the last three to five years in North America and Europe, including 'Governing Algorithms' (Barocas *et al.* 2013) and the one that gave rise to this book project (Ruhe 2014). Together, these different perspectives have raised crucial epistemological questions as to what would constitute the most appropriate scope for studying algorithms. For instance, what would be too narrow or too broad? And what constitutes the ideal distance to study algorithmic culture, allowing for a critical reflexivity without being too detached or removed from the actual practice and operation of algorithms? To this can be added the problems often associated with so-called 'hot topics,' that is, the pursuit of the 'new' for its own sake, and how to avoid falling into the "trap of newness" (Beer 2013, 6–7; Savage 2009).

Conceptual innovation, in light of such questions and problems, might very well mean returning to, and relying and building on older but more solid foundations, which do in fact exist. What we propose in this introduction is thus to revisit and modify Alexander R. Galloway's classic intervention, which construes ours as an age of algorithmic culture (2006). This idea of culture as marked by the algorithmic resonates strongly with the encompassing yet established discipline of cultural sociology and its efforts 'to take meaning seriously,' i.e., to understand 'meaning' not as a series of intangible or untethered significations, but as something deeply rooted in reality, agency, and performativity. Indeed, a cultural sociology of the algorithm is possible only insofar as algorithms are considered as both meaningful *and* performative, that is to say, performative *for the very reason* that they are meaningful, and vice versa. It is our contention here that while the aforementioned perspectives are all significant contributions, they generate rather than obviate the need for thicker, deeper, and more complex analyses of the kind of culture that algorithms are currently shaping. As the title of this volume suggests, we want to engage with this possibility of an algorithmic culture by supplementing or contaminating it with observations on pluralization.

The plurality of cultures in algorithmic cultures

Despite its theoretical potency, Galloway's innovation was never fully developed, and appears more inspirational than analytical. Of late, it is mostly Ted Striphas who has led what he calls "historico-definitional" efforts in determining what could more fully constitute such an algorithmic culture (2015, 2009; Hallinan and Striphas 2014; see also Roberge and Melançon forthcoming; and to a lesser extent, Kushner 2013). And the way he puts things in perspective has a rather humanistic tone: "What does *culture* mean, and what might it be coming to mean, given the growing presence of algorithmic [recommendation] systems [...]?" (Hallinan and Striphas 2014, 119). His attempt, in other words, is geared towards finding essential, if not ontological, categories under the terms "work of culture" or "world's cultural heritage," and their fundamental transformation through automation. For Striphas, it is all of the circulation, sorting, and classifying processes that are now dictated by "a court of algorithmic appeal." This too is a powerful notion; Striphas's argument is worth mentioning as it is epistemologically sound and captures the important stakes in this debate. On the one hand, he never fails to acknowledge the dual nature of algorithmic culture, or the way its semantic dimensions are inseparable from its more technical ones. On the other hand, he fully appreciates how the very 'publicness' of culture is currently being black-boxed through processes of privatization, to which we return below. The problem, small as it is, is elsewhere. If Striphas's arguments can be criticized at all, then it will be for their tendency to be relatively abstract and broad. To say that we are witnessing a shift towards algorithmic culture does not necessarily have to be an all-encompassing theoretical move. His idea of algorithmic culture remains *one* concept of *one* culture. In the end, as much as it is meaningful and consistent, it struggles to recognize the variety of algorithms today, and the ways they are fractal and heteronomous by definition. So how do we proceed from here? How can we develop an understanding of algorithmic culture that takes meaning seriously by being especially attentive to its inherent performativity and messiness? One possible way is to go even further back in time, to another seminal author who preceded Striphas and Galloway. In the 1970s Michel de Certeau wrote *La culture au pluriel*, in which he insists that any definition of culture would have to conceive of it as *un multiple* (1974; translated by Conley as *Culture in the Plural*, 1998). While he could not have been aware of the significance algorithms would later gain, his idea is nonetheless vital, and inspirational in this context. Indeed we are currently living in the age of algorithmic *cultures*.

Although difficult to represent in simple logical terms, one thing can be many, and multiple disparate things can be very commensurable. Such is an archipelago—for instance, the Bahamas and the Philippines—to give a metaphorical example. In the case of algorithmic cultures, it is necessary to make sense of how a certain enclosure is nonetheless part of a larger whole. There are of course many ways to explain such an enclosure; one that has become almost mainstream in cultural sociology comes from the Yale School, which insists on giving cultural realities a 'relative autonomy' in the way their terms are often dependent on one another (see Alexander 2004, 1990; Alexander and Smith 2002, 1998; see also Sanz and Stančík 2013). As for algorithms themselves, they develop a routinized 'inside,' an internal or auto-referential logic that is all interrelated meanings. They are a textual reality even before they are mathematical calculations; they crystallize imaginaries, hopes, expectations, etc. As Valentin Rauer puts it later in this volume, "Algorithms are part of a broader array of performativities that includes, for example, rituals, narratives, and symbolic experiences" (p. 142). As contingent normalizers and stabilizers, they have a symbolic life of their own which, like texts, only makes sense in a particular context. Cultural sociology rests here on what may constitute an original, yet very solid theoretical ground. Jeffrey Alexander's notion of "relative autonomy" resonates with Lorraine Daston's more recent narratological perspective, for instance, which inquires into the specific "history and mythology [...] of the algorithm" (2004, 362). To give a concrete example of how an algorithm, or a set of algorithms—a network or a specific family, so to speak—develops *by, of,* and *for* its own, our contributor Lucas Introna has shown elsewhere how algorithms used to detect plagiarism also alter the long established definition of what it means to produce an 'original' text. As algorithms can identify matching copies by fastening upon suspicious chains of words, writers have adapted their style of writing. Plagiarism algorithms are thus only able to detect "the difference between skillful copiers and unskillful copiers," and thereby performatively and somehow paradoxically produce the skillful copier as an 'original' author, resulting in an entire culture surrounding the sale of 'original' essays and ghost-writing services (Introna 2016, 36). Hence, instead of treating algorithms as mere utilitarian devices, the study of algorithmic cultures rather identifies the *meaningfully performative* effects that accompany algorithmic access to the world: What is it that they *do*, culturally speaking? How do they *make sense* of their surroundings and the different categories people use to interpret them?

As it turns out, one of the most salient points to be made in this introduction revolves around algorithmic cultures as being *un multiple*. Nick Seaver offers a similar argument when he notes that "rather than thinking of algorithms-in-the-wild as singular objects, [...] perhaps we should start thinking of them as a population to be sampled" (2014, 6). Algorithms are dynamic entities that mesh with specific sets of knowledge and experience in textured and complex ways. Thus, another appealing way to make sense of their relative autonomy and enclosure is to borrow from the language of cybernetics (Totaro and Ninno 2014; Becker 2009). Feedback loops, decision-making by classification, continual adaption, and the exchange of information are all characteristics of recursive quasi-circular routines that typify the non-linear unfolding of algorithms, as seen above. Göran Bolin and Jonas Andersson Schwartz have recently given this idea a practical spin, noting that

> (a.) in their daily operation, professionals have to anticipate what the end-user will think and feel; [... and that] (b.) many everyday users try to anticipate what the [...] media design will do to them, [...] which involves a recourse back to (a.)
>
> (2015, 8)

Google could serve as a prime example here. Complex and multivalent, there exists, as our collaborator Dominique Cardon calls it, something like a unique "PageRank spirit" (2013; see also in this volume), in which symbolic as well as performative aspects are constantly interacting. Such a spirit is easy to spot in the cyclical anticipation of needs, the satisfaction of experience, and the personalization of navigation, all typical of the contemporary search engine. It is also evident in the implementation of sophisticated algorithms over the years—such as Panda, Penguin, Hummingbird, and Pigeon—and how they have helped in the on-going struggle against the polluting power of search engine optimization (see Röhle 2009). Lastly, this particular spirit is present in how Google has tried to find a balance between its sense of natural, meritocratic indexing and its own commercial needs, which then serve to subsidize its more futuristic technological endeavors. Not only are these three examples recursive in themselves, but they also end up swirling together and influencing one another to create a distinctive, powerful, and meaningful algorithmic culture. This is precisely Google's own "culture of search" (Hillis *et al.* 2013) or, to put it more bluntly, the "Googleplex" (Levy 2011). Is this to say that the company has no sense of what is going on outside? Certainly not. Rather, this particular culture can co-operate with others, and may even coincide with others in many respects, but it does not mean our analysis should conflate them all. A finer understanding of algorithmic cultures, in other words, should be able to zoom in and zoom out, to see the particularities of each algorithmic culture, as much as what they also have in common.

Examples of this abound: individuality *and* reaching, particularity *and* sharing, distinctiveness *and* commensurability, small *and* big picture. For algorithmic cultures can of course cut across various social, economic, and political spheres; for instance, when a particular usage of predictive algorithms in the stock market borrows its probabilistic methods from games of chance, transporting them into another field, and thereby transforming them for its own practical needs. Or when developments in artificial intelligence are derived from computer algorithms in the game of chess, thereby shaping the very future of artificial intelligence for years to come (Ensmenger 2012). Thus, algorithmic cultures are not based on a fixed and unmoving ground, but are rather more like mobile methods that are adapted, transformed and made to measure for each particular use. In fact, this entire volume serves as proof for this argument. Each chapter develops a unique take on what it means for algorithms to be culturally entrenched and performative; each of them explores the density extending from a particular assemblage or ecology by proposing a specific interpretation. The exact description of the chapters' contents will come in a moment, but suffice now to say that it also falls on the reader to navigate between them, to ask the questions s/he judges appropriate, and to wrestle with the different intellectual possibilities that are opened up.

To argue that algorithmic cultures are *un multiple* still opens, rather than forecloses, the need to find a plausible solution to the problem of what could constitute their variable yet common nature. There must be something; indeed, algorithms revolve around a question or an issue that is each and every time particular but nonetheless always similar. We want to suggest here, as others have, that such important stakes constantly bring about and thus recycle "the power to enable and assign meaningfulness" (Langlois quoted in this volume in Gillespie 2014; see also Roberge and Melançon forthcoming). This is a question as old as the idea of culture itself, and the social sciences have been aware of it for their entire existence too, from the moment of their founding until today (Johnson *et al.* 2006). Culture needs legitimacy, just as algorithms and algorithmic cultures need legitimacy. It is about authority and trust; it is about the constant intertwining of symbolic representation and more prosaic performance, the production as well as the reception of discursive work. In our current day and age, we are witnessing the elaboration of a kind of 'new normal' in which algorithms have come to make sense in the broader imaginary; they are 'accepted' not because they refer to something transcendent in the classical sense, but because they have developed such acceptability in a newer, more immanent way. Scott Lash's insight regarding algorithms' principle of "legitimation through performance" is fundamental in this regard (2007, 67). In their actual real-time unfolding, algorithms implicitly or explicitly claim not only that they are cost-effective, but moreover *objective*, in both an epistemological and a moral sense. Again, this occurs in a very mundane way; their justification works, as much as it is rooted in an enclosed routine that says very little in fact: *algorithms work straightforwardly, they provide solutions, etc.* Neutrality and impartiality are whispered and tacitly assumed. Tarleton Gillespie notes something similar when he underscores that "more than mere tools, algorithms are also stabilizers of trust, practical and symbolical assurances that their evaluations are fair and accurate, free from subjectivity, error, or attempts at influence" (Gillespie 2014, 179; see also Mager 2012). That is the magic of something non-magical. Objectivity as an information process, a result, and a belief is the equivalent of legitimacy as the result of a form of belief. The strength of algorithms is their ability to project such objectivity to the outside world (to what is in their rankings, for instance), while accumulating it 'inside' the algorithms themselves as well. This is because any provider of value *ought to be* constructed in a way that is itself valued. Gillespie is astute on this point, noting that "the legitimacy of these functioning mechanisms must be performed alongside the provision of information itself" (2014, 179). Here legitimacy acquires an ontological dimension.

This is not to say that the quest for legitimacy is an easy endeavor—quite the contrary. Performance and justification exist only insofar as they can find an audience, to the point in fact where the 'reception' part of the equation is just as important. The problem, of course, is that such reception is inherently cultural and constituted by interpretation, expectation, affect, speculation, and the like (Galloway 2013; Seyfert 2012; Kinsley 2010). Reception, in other words, is unstable and uneven by its very definition. What Lash calls "legitimation through performance" is for this reason nothing less than a steady negotiation—in terms close to those discussed above. Performance and reception interweave in such a way as to constitute specific routines and cultures in which the trust afforded to algorithms cannot foreclose the possibility of contestation. The hopes and desires of some could very well be the fears and dislikes of others. And while justification is performative, so too is criticism. The controversy that erupted around Google Glass is a case in point. Research into their Glass Explorer program initiated by one of us has indicated how much style and design has been figured into the corporate planning for wearable computing (Roberge and Melançon forthcoming). For example, to give Google Glass a broader appeal, the company hired a Swedish designer to help design the device, including its color palette and minimalistic contours (Miller 2013; Wasik 2013). Regardless, the critical response was negative, noting that Glass is "so goddam weird-looking," "ugly and awkward," and makes interaction "screamingly uncomfortable" (Honan 2013; Pogue 2013). Social and cultural discomfort with this new form of interaction helps explain the algorithmic device's critical reception. In the end, it was the pejorative term "glasshole," symptomatically blending aesthetic and normative-moral judgments, that proved one of the most influential factors that forced Google to withdraw. What this example thus shows is how ambiguous various meanings and interpretive conflicts, as well as the algorithmic cultures they shape, end up being. Messiness is not an option; it is an ongoing and transformative characteristic.

Algorithmic traffic: calculative recommendation, visibility and circulation

The key idea behind this volume on algorithmic cultures is that such cultures are plural, commensurable, and meaningfully performative. The purpose here is to offer a "thick description" *à la* Geertz (1973), i.e., an analysis of different routinized unfoldings that revolve around rich and complex stakes and issues. Legitimacy is certainly one of these. Everyday life is full of occasions where this question is not raised, but here the stakes are tremendous, as they encroach on some sort of cultural core. Algorithms are sorters; they are now key players in the gatekeeping mechanisms of our time (Hargittai 2000). To be sure, gatekeeping has been around for a long time, from the arts patrons of the classical age to modern-day newspaper critics. But this only strengthens the argument: the role played today by algorithms still adheres to a prescriptive selection of ascribing value, for a particular audience, with all of the attendant moral and political valences. Gatekeeping is about making editorial choices that others will have to deal with. It is about taste and preference-making, which explains, at least in part, why many recommendation algorithms are so influential today, from Amazon to Netflix, YouTube, and the like. Beer synthetizes this point nicely:

> It is about the visibility of culture, and of *particular forms of culture that algorithmically finds its audience*. These systems shape cultural encounters and cultural landscapes. They also often act and make taste visible. The question this creates is about the power of algorithms in culture and, more specifically, the power of algorithms in the formation of tastes and preferences.
>
> (Beer 2013, 97, emphasis added)

Two recent articles in particular have captured this trend and how it has evolved in specific settings, one in terms of film (Hallinan and Striphas 2014), and the other in music (Morris 2015). Netflix, and specifically the Netflix Prize, is emblematic in many regards; launched in 2006, the contest offered US$1 million to whoever could first boost the accuracy of their recommendation algorithm over the benchmark of 10 percent. As the challenge was a huge success among computer scientists in the U.S. and abroad, it represents for Blake Hallinan and Striphas a prime example of how "questions of cultural authority are being displaced significantly into the realm of technique and engineering" (2014, 122). Yet this is only one half of the equation. The other half deals with the logic or the economic purpose enabling such a quest for personalized recommendation, something the authors call a "closed commercial loop," in which "the production of sophisticated recommendation produces greater customer satisfaction which produces more customer data which in turn produce more sophisticated recommendations, and so on" (122). Where information processing becomes key, the meaning of culture drifts toward simpler views on data, data-mining, and the value it produces. This is what Jeremy Wade Morris finds as well in his study of Echo Nest, the "taste profiling" platform acquired by the music streaming service Spotify in 2014. The management of massive databases and new behavioral tracking techniques, by those that Morris calls "infomediaries," now relies "on the efficacy of the algorithms [...] to know what is essential about you and your tastes" (2015, 456). This is the case because it essentially opens the door to "highly segmented and targeted advertising opportunities" (455). This logic or trend is indeed very strong, though it is not the only one at play. Morris's argument is subtle enough to recognize the pervasiveness of human-maintained playlists as a mode of alternative curation that most of today's platforms are unable to let go of. These human-to-human taste dialogues, so to speak, still exist in most music streaming services as a way to cope with the abundance of content. Both automated and so-called 'manual' gatekeeping mechanisms thus co-exist more or less side by side in a sort of complex, if tacit and very delicate, tension.

The data-intensive economy and culture that is currently taking shape is also of interest to Lucas Introna in his contribution to our volume. By tracing the genealogy of online advertising, he analyzes recent forms of what he calls "algorithmic choreography." While traditional online advertisements indiscriminately place ads on sites that all users will encounter—a banner on the top of a webpage, for instance—more innovative brokers such as Dstillery adapt to what they perceive as the needs of the individual. Data-mining, behavioral targeting, contextual advertising, machine-learning algorithms, and the like are thus all part of the same arsenal. The aim here is finding a "market of one," where particular subjects are addressed through personalized advertisements. Time and again, it is about addressing "the right person at the right time with the right creative content" (p. 41). Such a choreography requires and enacts particular forms of subjectivity, which Introna calls "impressionable subjects," i.e., subjects that are willing to be impressed by the information the algorithm has prepared for it at any given moment. In one way of reaching customers in an online advertisement called "prospecting," data are collected from user activities on the spot (through clicks, queries, etc.). From such data, correlations can be derived and users can be "branded": whoever visits a particular page, for example, might be interested in the same products as another user who visited similar sites. On the one hand, in algorithmic cultures the subject is treated as a mere statistical entity, a branded subject. On the other, subjects are not entirely passive, but rather are actively engaged in the selection of information they see and how they are shaped by it; they partially curate what they are going to see (and perhaps buy) through their own behavior. Thus, user behavior and online advertising become deeply cultural and social affairs because they either enact subjects or fail to connect with them. Introna shows how in their own way algorithmic cultures are *un multiple*, that is, very generic but at the same time very personal. Placing an advertisement correctly enacts or confirms the subject in a highly personalized way: who I am becoming depends on where I am surfing. In turn, incorrectly placing an advertisement is not only a missed opportunity, but can also question and insult the subject ('Why am I seeing this?').

In his contribution, Tarleton Gillespie investigates the complexity and heterogeneity of automated gatekeeping by addressing the rich yet understudied subcategory of trending algorithms. Indeed, these are everywhere today, from Buzzfeed to Facebook and Twitter; they are an icon of a new genre that is oftentimes the icon of themselves, since "trending is itself trending." Gillespie's fine-grained analysis thus starts by asking not what algorithms do to cultural artifacts, but instead "what happens when algorithms get taken up as culture, when their kinds of claims become legible, meaningful and contested" (p. 69). Such algorithms appear as a measurement ritual, but of exactly what is less clear. Is it a glimpse into the popularity of different content, as was *American Top 40* or *Billboard*? Is it a small window into 'us,' with the attendant problem of defining exactly who this 'us' is—a public, a nation, etc.? Or is it simply about capturing some sort of pulse, velocity and movement in between undisclosed and thus incalculable points? Surprisingly, all these difficulties are fueling, rather than extinguishing, the urge to measure and position measurement as a meaningful accomplishment. In other words, trending algorithms are popular because they are inherently ambiguous. In addition, real and practical biases are numerous, as if they were inscribed in the very DNA of these algorithms. According to Gillespie, this has to do with the black box character of most social media platforms. More important, however, is the fact that biases are above all interpretations of biases, in the way that they depend on the expectations, hopes, and desires of those who care enough. Validity is a cultural question in this regard. For instance, many have criticized Twitter and Facebook for the triviality of their trends, while at the same time often underscoring that their own favorite 'hot topic' was not appearing. Controversies related to trending algorithms are simply not about to vanish. They emerge from time to time, depending on different places, people and issues, as a symptom of something deeper—indicating a fundamental conflict over legitimacy.

Gatekeeping, as has become clear, represents an issue with both representational and performative ramifications. As it deals with the visibility and circulation of pretty much everything cultural, it has been fundamentally transformed by the dissemination of algorithms. The challenge to the authority-thrust nexus of all gatekeeping mechanisms is thus as significant as those mechanisms are constant. For the social sciences, too, this represents a substantial challenge, one that forces us to develop new holistic understandings as well as new and more empirical analyses (Kitchin 2014; see also Ruppert *et al.* 2013). In their contribution to this volume, Jean-Samuel Beuscart and Kevin Mellet offer an excellent example of the latter. They study *LaFourchette.fr* and other consumer rating and review sites as a now more-or-less standardized, if not ubiquitous, tool on the Web. What their findings show, however, is that the massive presence of such platforms is not antithetical to a sense of agency among users, and that the latter has given rise to a rich and interesting negotiation among actors, both human and non-human alike. Frequent writers of reviews, for instance, are indeed moved by a non-negligible dose of reflexivity. According to Beuscart and Mellet, "at least part of the effectiveness of this phenomenon is the ability of users to build a coherent pattern of use that regulates their evaluation behavior to work towards a collective aim" (p. 90). Self-esteem thus derives from a sense that somehow there exists a form of readership that also forms a rational and socialized judgment. This might create a distant image of what constitutes a collective intelligence, and such an image is active enough to be considered performative.

Not to be forgotten is the question of whether the actual fragmented nature of recommendation algorithms constitutes *un multiple*. Different calculation routines clearly produce different outcomes, and from there it becomes important to assess what this could mean, both ontologically and epistemologically. Putting things in such a perspective is the task Dominique Cardon sets for himself in his contribution to our volume. He proposes, in essence, a classification of classificatory principles, focusing on the ways that they are not simply and straightforwardly dependent on economic forces, but also on one another, by way of relation, opposition, comparison, etc.—a conceptual move closely linked with Alexander's "relative autonomy of culture," as seen above. Cardon discusses four types of calculation and the ways they inform the "competition over the best way to rank information": *beside the Web*, as a calculation of views and audience measurement; *above the Web*, as a meritocratic evaluation of links; *within the Web*, as a measure of likes and popularity; and finally, *below the Web*, as the recording of behavioral traces that allows for more tailored advertising. These four types reveal very different metrics, principles, and populations to be sampled, and yet they are commensurable in that together they inform a "systemic shift" in how society represents itself. "Digital algorithms," writes Cardon, "prefer to capture events (clicks, purchases, interactions, etc.), which they record on the fly to compare to other events, without having to make broad categorizations" (p. 106). Statistics as we used to know them, such as those relying on large variables like sex and race, are being replaced with more precise and individualized measurements. In turn, society appears as an increasingly heterogeneous ex-post reality, the best explanation of which might be that there is no real, fundamental, or comprehensive explanation—with all the consequences that this entails for the social sciences.

From algorithmic performances to algorithmic failures

Instability, fragility and messiness all gesture at the praxiological character of algorithmic cultures. In contrast to the dominant paradigm of computer science, which describes algorithms as procedural and abstract methods, we conceptualize algorithms as practical unfoldings (Reckwitz 2002). Galloway, in his seminal essay, already points to the *pragmatic aspect* of algorithmic cultures: "to live today is to know how to use menus" (Galloway 2006, 17). As users, when we operate in algorithmic cultures, we operate algorithms. For instance, the handling of software menus is a practice (interactions and operations with others, human and non-human alike) in which we manage algorithmic devices: we schedule meetings on our online calendar, set up notifications on emails, program our navigational devices to lead us home, etc. We activate and deactivate algorithms to govern our daily life. Thus, algorithms are not so much codes as they are *realizations of social relations* between various actors and actants.

As practices, algorithms are distinguished by recursive and very entrenched *routines*. Algorithms are supposed to help in the performance of repetitious tasks; they implement activities for reduced cognitive and affective investment, and thereby make it possible to focus on more important and perhaps more interesting tasks. The analysis of algorithms as routines (or routine practices) accounts for deviations from the mathematical and technical scripts, deviations that emerge from various sources, such as a failure in design, incomplete implementation, and the messiness of operations or interactive effects between different algorithmic and non-algorithmic actants. This is something computer science can barely do, as it is in its DNA, so to speak, to define algorithms through *precision* and *correctness*. Computer scientists accept deviations only in human routines, and thus foreclose the possibility that not every repetition is identical; rather, each iteration of the routine introduces little deviations in each step (Deleuze 1994). We would even go so far as to say the discourse of the discipline of computer science conceptually excludes algorithmic practices, and hence the possibility of their deviations from the script. For cultural sociology, the assignation of deviations exclusively to humans seems problematic. The notion of an algorithmic precision and correctness seems to be rather part of the tale of an *algorithmic objectivity* discussed above, a quest for a higher rationality, where algorithms act autonomously and supercede human routines. In this tale, algorithms promise an identical repetition that allows for easy modeling and precise predictions. However, such *imaginaries* of algorithmic cultures, their promises and dreams, have to be distinguished from algorithms in practice.

In algorithmic cultures, we witness changes of social relations, for instance the emergence of highly customized relations. In Joseph Klett's contribution to this volume, he gives an example of the transition from digital stereo to "immersive audio" that exemplifies such a change. Stereo sound (the sound we get from traditional stereo speaker systems) operates with generic relations: each audio speaker establishes a fixed relation to a 'user,' which really is an invariant sensory apparatus located in a fixed point in space (the so-called 'sweet-spot').

In contrast, relations in algorithmically realized soundscapes are highly personalized. Klett shows how audio engineering, as with many other technological apparatuses, is moving from the use of algorithms as general mediators to the use of algorithms as highly specific mediators between technological devices and singular individuals. Such personalization allows for a much richer audio experience, because we do not have to find the optimal spot of sound exposure; instead, the sound is continuously adapting to our individual perspective. Inevitably, the transition from generic relations to dynamical adaptive relations through algorithms has consequences for social life. By adapting to individual bodies and subjects, personalization algorithms also change the very nature of social relations, disentangling and cutting off some relations and creating new ones. Personalization algorithms in noise-cancelling headphones are an example of such disconnections; they deprive social relations of acoustic communication. Thus, personalization algorithms create enclosures around the subjects where "the body becomes a part of the audio system" (p. 116). Together, body and device create a closed algorithmic culture.

In this day and age, algorithmic relations are not only enacted by and with humans, but also by and with algorithms themselves. There are indeed endless chains of algorithms governing one other. Understanding such relations will cast doubt upon the purported antagonism between humans and computer algorithms, between humans and algorithmic routines—antagonisms endemic to the proposals of computer science, approaches that generate notions like algorithmic objectivity and pure rationality. The crafted imaginary that reiterates and relies on the classic myth of a struggle between man and machine (as exemplified in mythical events such as Kasparov vs. Deep Blue) ignores human immersion in algorithms (such as the programmers' immersion in Deep Blue—their tweaking of the programming between matches to adjust to Kasparov's play). It bears repeating that the definition of algorithms as formal procedures focuses only on precise and identically repeatable processes, while the examination of practices and performances takes into account deviations and divergences. Unstable negotiations, slippage, fragility, and a proneness to failure are in fact important features of algorithmic cultures. In 'real life,' algorithms very often fail, their interactions and operations are messy. This is particularly true when they tumble in a sort of in-betweenness among other actors (algorithmic or not), where they tend to deviate from their initial aim as much as any other actant.

The emergence of failures has to do with the complexity of interactions. Interactions that are not only face-to-face or face-to-screen, but that also take place within complex assemblages, contribute to the production of errors and bugs. Countless examples of such failures can be found, from the (mis)pricing of "Amazon's $23,698,655.93 book about flies" (Eisen 2011), to the demise of *Knight Capital*, an algorithmic trading company that lost about US$400 million in a matter of 45 minutes due to a malfunctioning trading algorithm (SEC 2013, 6). Consequently, the everyday use of algorithms results in a mixture of surprise and disappointment. The astonishment often expressed when Amazon's recommendation algorithms correctly predict (or produce) our taste, and directly result in a purchase, goes hand in hand with complaints of how wildly off the mark they are. We have come to expect failing algorithmic systems and we have indeed become accustomed to dealing with them. Making fun of such failures has become a genre in itself: "@Amazon's algorithms are so advanced, I've been offered over 10,000 #PrimeDay deals and am not interested in any of them" (Davis 2015).

In his contribution to our volume, Shintaro Miyazaki explains the avalanching effect of "micro-failures" in algorithmic cultures. He shows how something that might seem miniscule, irrelevant, a small divergence in code, an almost indiscernible misalignment, can be leveraged to catastrophic results in algorithmic feedback processes. Miyazaki's historical case study of the AT&T Crash from 1990 shows that such failures have been part of algorithmic cultures from very early on. In this case, a software update in AT&T's telephone network created a feedback loop in which the entire system created an unstable condition from which it was not able to recover. While separate subsystems contained emergency routines that enabled each to automatically recover from cases of malfunction, the algorithmic feedback loops across subsystems caused interacting algorithms to turn one another off. This resulted in an algorithmic network with unproductive operations, which stem from what Miyazaki calls "distributed dysfunctionalities" (p. 130).

If we were to take seriously the fact that failure is an inevitable part of algorithmic life, then Miyazaki's analysis of "distributed dysfunctionality" has a further implication—namely, that distributed dysfunctionality may in fact be a process where a network of algorithms inadvertently creates a higher form of an *ultimate machine*. The prototypical ultimate machine was created by Claude E. Shannon. It has one, and only one, particular purpose—to turn itself off:

> Nothing could look simpler. It is merely a small wooden casket the size and shape of a cigar-box, with a single switch on one face. When you throw the switch, there is an angry, purposeful buzzing. The lid slowly rises, and from beneath it emerges a hand. The hand reaches down, turns the switch off, and retreats into the box. With the finality of a closing coffin, the lid snaps shut, the buzzing ceases, and peace reigns once more.
>
> (Clarke 1959, 159)

Because of its particular functionality, the ultimate machine was also named the *useless machine* or *leave me alone box*. The case described by Miyazaki may be understood as a more complex version of such a machine. In fact, it was not a single machine that turned itself off, but rather a chain of machines performing algorithmic interactions, so that each machine turned its neighbor off, right at the moment when the neighbor's recovery operation had been completed. While a simple ultimate machine still requires humans to flip the switch, algorithmically distributed dysfunctionality incorporates this function, creating a stable instability that requires non-algorithmic actors to end those dysfunctional and the non-productive routines. This is a case of an algorithmic practice where algorithms start to act and interact according to a pattern that had not been inscribed into them, making them essentially unproductive. One might describe such a machine as an algorithmic Bartleby, where the demand to initiate routines is countered by the algorithmic expression *I would prefer not to*. Such a description has perplexing explanatory value, especially if we contrast it with our earlier definitions of algorithms as routinized unfolding. As much as Bartleby's refusal affects the daily routines at work, algorithmic dysfunctionality also addresses those routines, undermining them and making them *unproductive*.

Cases of unstable algorithms are not unusual. In algorithmic trading, it is not uncommon for traders to have to force algorithms out of unstable conditions. For instance, software bugs or feedback loops might cause an algorithm to flicker around thresholds, where it continuously places and cancels orders, etc. (Seyfert forthcoming). Even though the phenomenon is very difficult to trace, some scholars have also argued that many unusual market events can be explained by such non-productive routines (Johnson *et al.* 2012; Cliff *et al.* 2011; Cliff and Nothrop 2011). To give an example, an initial analysis of the Flash Crash of 2010 suggested that such non-productive algorithmic interactions might have been the culprit. The Flash Crash describes a very rapid fall and consecutive recovery in security prices. The Joint Report by the Commodity Futures Trading Commission and the Security Exchange Commission in the United States described it in the following way:

> At about 2:40 in the afternoon of May 6, prices for both the E-Mini S&P 500 futures contract, and the SPY S&P 500 exchange traded fund, suddenly plunged 5% in just 5 minutes. More so, during the next 10 minutes they recovered from these losses. And it was during this recovery period that the prices of hundreds of individual equities and exchange traded funds plummeted to ridiculous levels of a penny or less before they too rebounded. By the end of the day everything was back to 'normal,' and thus the event was dubbed the May 6 Flash Crash.
>
> (CFTC and SEC 2010a, 3)

According to this Joint Report, high-frequency traders (relying on algorithms)

> began to quickly buy and then resell contracts to each other—generating a 'hot potato' volume effect as the same positions were rapidly passed back and forth. [...] HFTs traded over 27,000 contracts, which accounted for about 49 percent of the total trading volume, while buying only about 200 additional contracts net.
>
> (CFTC and SEC 2010a, 3)

This *hot potato effect* is another iteration of distributed dysfunctionality, an unproductive routine that inadvertently subverts the productivity paradigm of the financial markets.

One reason for the emergence of failures in algorithmic practices has to do with the fact that interactions with and among algorithms often tend to be misunderstood. In his contribution, Valentin Rauer shows in two case studies the problems in assessing algorithmic agency. In algorithmic cultures, traditional interactions through deictic gestures have been replaced by what Rauer calls "mobilizing algorithms." While face-to-face interactions allow for deictic gestures such as *this* or *you*, interactions over distance require intermediaries. Mobilizing algorithms have become such intermediaries, operating to a certain extent autonomously. Examples are automated emergency calls that serve as functional equivalents to deictic gestures (Mayday! Mayday!). Rauer shows that the introduction of such algorithmic intermediaries leads to varying scales and ranges in capacities to act. Such scaling processes make the notion of a purely algorithmic or human agency problematic. Self-sufficiency and complete independence are thresholds, or rather each constitutes a limit that is never fully reached in either humans or algorithms. But in public discourse, such scales of agency are ignored and obfuscated by strong imaginaries. The problems with these imaginaries become especially visible at the moment of algorithmic breakdowns. Rauer illustrates this with the case of a "missing algorithm" that ultimately led to the failure of the Euro Hawk drone project. In this particular circumstance, a missing algorithm caused the drone to fly on its first flight "unguided and completely blind, posing a real threat to anything in its vicinity" (p. 146). That particular algorithm was 'missing,' not as a result of an unintentional error, but rather, because the drone was supposed to be guided—that is, governed—by an acting human. Thus, the prototype of Euro Hawk operated with a strong notion of human agency—an agency that always masters its creations—while the agency of the drone was underestimated. The *missing algorithm* shows that failures and messiness are crucial to algorithmic practices.

Paradoxical as it seems, a missing algorithm is part of the messiness in algorithmic practices, a messiness that is also the reason for the promises and dreams inherent in algorithmic cultures. That is to say, the fulfillment of this dream is *always one step away* from its completion. There is always only one more algorithm yet to be implemented. In other words, it is only such constant algorithmic misalignments that explain the existence of promises and hopes of a smooth algorithmic functionality. If everything were functioning smoothly, these promises would be superfluous and would simply disappear. Strictly speaking, the dream of algorithmic objectivity, of smooth operations and efficiencies, of autonomy and the hope of a higher rationality, makes sense especially in contrast to constant failures.

Furthermore, misalignments and failures in algorithmic cultures are not only due to missing algorithms and bugs, but may precisely be attributable to the mismatch between the expectations of algorithmic rationality, agency, and objectivity inscribed in the codes on the one hand, and actual algorithmic practices on the other. When algorithms enter into socio-technical assemblages they become more than just "Logic + Control." Thus, a cultural analysis of algorithms cannot just include the technical niceties of codes and technical devices, i.e., their technical functionalities; it will also need to focus on the complex of material cultures, technological devices and practices. Hence, it is problematic when contemporary studies of algorithms primarily focus on the creepiness and suspicious nature of algorithms, which are hinted at in conference titles such as "The Tyranny of Algorithms" (Washington, December 2015). Such perspectives not only ignore the very mundane nature of the disappointments caused by algorithms but also the logical dynamics between promise and disappointment operating in algorithmic cultures. These studies tend to conflate the industries' imaginaries of rationality, autonomy, and objectivity with actual practices. They (mis)take the promises of those who construct and, most importantly, sell these systems for the realities of algorithmic cultures. Where they should be analyzing the 'legitimation through performance' of algorithmic cultures, they end up criticizing imaginaries and their effects, irrespective of the praxiological processes of actualization (or non-realization) of these imaginaries. In their preferred mode of criticism they fall prey to what Mark Nunes has called "a cybernetic ideology driven by dreams of an error-free world of 100 percent efficiency, accuracy, and predictability" (2011, 3). Consequently, by overestimating the effectiveness and by ignoring the messiness and dysfunctionality of algorithmic practices, these cultural and social analyses take on the character of conspiracy theories in which "secret algorithms control money and information" (Pasquale 2015).

The rather conspiratorial attitudes towards algorithms might also be explained by the sheer magnitude of the ambiguity that is involved in algorithmic cultures. Algorithmic practices, where we use and where we are being used by algorithms, involve tacit knowledge. Most of us use algorithms every day, we govern them every day, and we are governed by them every day. Yet most of us do not know much about the algorithmic codes of which these algorithmic assemblages are made. This non-knowledge makes us suspect something uncanny behind the screen, something that is fundamentally different from the intentions of our human companions. It is the lack of information that leads some human actors to ascribe intentions to all algorithmic activities, a general attitude of suspicion that Nathalie Heinich has called the "intentionalist hypothesis," that is, a "systematic reduction of all actions to a conscious (but preferably hidden and thus mean) intention" (Heinich 2009, 35). It is this ambiguity that makes the analysis of algorithmic cultures in social and cultural studies particularly germane. The production, usage, and failure of algorithmic systems are stabilized by cultural narratives that resort to powerful imaginary expectations. Thus, in order to see this tension between practices and imaginaries, to grasp algorithmic cultures in their constitutive tension, it is not enough to focus on the cultural narratives of those who explain and promote algorithmic systems and on those who express conspiratorial fears: focus on the algorithmic practices themselves is also required, for it is here where failures are most visible.

Cultivating algorithmic ambiguity

Because algorithmic circuits are interactions between very different human and non-human actors, they are ambiguous, and it becomes particularly difficult to locate agency and responsibility. Consequently, algorithmic circuits and interactions present a challenge, not only to the scholars in social sciences and cultural studies. Interpretations vary widely, and the distribution of agency and the attribution of responsibility shifts, depending on the epistemic formations of the interpreters of particular events. While some authors like Miyazaki focus on pure algorithmic interactions (Miyazaki [in this volume]; MacKenzie 2015; Knorr Cetina 2013), others conceive of them as distributed functionality between humans and algorithms, as "blended automation" (Beunza and Millo 2015), while some even go so far as to see in algorithms nothing but instruments of human agency (Reichertz 2013). Political systems especially tend to resort to the last view, in particular when things go wrong and accountable actors need to be named. Here, the Flash Crash of 2010 and its interpretation by the Security Exchange Commission in the United States is a particularly apt example. The rapidity of the fall in stock market prices and their subsequent recovery led to fingers being pointed at the interactions of trading algorithms of high-frequency traders. Early interpretations especially took this event as a new phenomenon, an event resulting from the interaction of complex technological systems ('hot potato effects'). However, as time went by, human rather than algorithmic agency was increasingly deemed accountable. A comparison between the first report of the Flash Crash by the CFTC and SEC from May 18 (CFTC and SEC 2010a) and the second report from September 30 (CFTC and SEC 2010b) shows an increasing focus on the inclusion of individual actors and their intentions. While the first report also includes the possibility of inter-algorithmic feedback loops (the aforementioned 'hot potato effects'), the most recent report from 2015 does not mention algorithmic interactions or any type of complex feedback loops. Instead, it points to a human trader, London-based Navinder Singh Sarao, who was the single individual actor named as being connected to the event (CFTC 2015a and b). Such reductionist explanations are highly contested within the field. For some, it seems highly improbable that a single trader can intentionally create such an impact on a trillion-dollar market (Pirrong 2015). If his activities did indeed contribute to the Flash Crash, then, it has been argued, it was rather as an unintentional butterfly effect, as conceptualized in complexity theory (Foresight 2012, 71–72).

However, as this example of the slow transition from blaming algorithmic interactions to blaming human intentions shows, the interpretation of algorithmic failures greatly depends on the epistemic paradigm used by the interpreter. That is to say, each interpretation stems from a particular way of sense-making, which includes the devices used to access an event. While information science, media studies, and STS have no problems ascribing agency, responsibility, and accountability to emergent phenomena stemming from inter-algorithmic events, the same is not true for political systems (or market authorities for that matter) that (still) tie responsibility to human actors. It is safe to say that the political system itself created the pressure on the SEC and CFTC to present an accountable actor with which traditional juridical systems can operate. Algorithms are certainly not (yet) among those. As we have seen, the emergence of algorithmic cultures is also accompanied by the blurring of clearly defined flows, creating an atmosphere of uncertainty about the identity of interactional partners.

Thus, one of the most important questions within algorithmic cultures is always "who we are speaking to" (Gillespie 2014, 192). In all types of social media platforms, the user needs to trust that s/he is interacting with an 'actual' user. That is especially important for economic interests, which rely on an unambiguous identification of senders and receivers of financial transmissions. Economic operations rest upon clear definitions of the party to whom (or which) we are speaking, for it is only then that we know the identities of those from whom we are buying or to whom we are selling.

In his contribution to this volume, Oliver Leistert shows that social media platforms solve this problem by operating with purification practices, which seek to ensure that our crucial communications are with 'real' users and real users alone. In turn, users need to believe that their counterparts are real, ergo, they need to *trust* the social media platform they are using. Thus, the "algorithmic production of trust" (p. 159) is one of the most important mechanisms of social media platforms. This is what such platforms actually *do*: rely heavily on trust to solve the problem of uncertainty. Leistert further describes the doubling mechanisms in conditions of uncertainty, where certain social bots are designed to *exploit the trust* that social media platforms painstakingly try to establish. He sees such social bots as machines that parasitically feed on our desires to be followed, to be ranked, and to be trending. As 'algorithmic pirates' they feed in various ways on 'pure' interactions. These desires can be exploited, for instance by the offer to 'automatically' feed it with fake followers, with bots that pretend to be 'real' followers. In addition, it is not uncommon for some—often commercial—users to buy followers on social media platforms. Another example is *harvesters* that attempt to friend as many users possible in order to extract user data. Not only do they feed on the desire of a particular user to enhance his/her popularity (through the increase in the numbers of followers), they also feed on the data flows that constitute the core business of social media platforms. Leistert hence describes real performative effects in algorithmic cultures. Not only is the general uncertainty regarding whom we are addressing exploited, the exploitation in fact increases uncertainty, even for bots. For instance, when 'social bots' mimic human users they increase uncertainty to the extent that they themselves become unsure whether or not they are still dealing with 'normal' users. Thus, bots themselves have to identify fake counterparts. On the one hand, algorithmic parasites pollute the pure interactions between 'normal' users that social media platforms try so hard to establish. But on the other hand, they too need to purify the pollutions their own actions have caused. In turn, what Leistert shows is how purification practices and parasitic bots performatively intensify and escalate the process of producing and reducing uncertainty.

The interpretations of algorithmic cultures are not just epistemic problems, questions of who is right or wrong. Where computer science defines algorithms as procedures or recipes for solving problems, approaches such as cultural sociology emphasize their performative effects, their recursive functions by which algorithmic practices not only create new problems, but also create the problems for which they are ultimately the answer. The performativity of algorithms is also (recursively) related to reflections in social and cultural studies itself.

Barocas and Nissenbaum (2014) have shown that the use of new technologies can initiate a reflexive process that helps us clarify already existing ideas. For instance, algorithmic practices do not simply, as is often suggested, challenge traditional notions of privacy, for instance in the context of Edward Snowden's revelations. Algorithmic practices such as Big Data do not simply threaten classic notions of individual privacy and anonymity, since they do not operate with classical features such as name, address, and birth place. Rather, they change the very definitions of what it means to be private and anonymous. By assembling algorithmic portfolios of the users they are tracing, they operate with entirely different features of their users, and thereby create new identities. Consequently, Facebook's shadow profile and what Google has rather cynically called our "anonymous identifier" (AdID) are effectively mechanisms in identity politics (Barocas and Nissenbaum 2014, 52–53). "Anonymous identifier" clearly differs from a classical identifier, in which identity corresponds clearly to names, addresses, social security numbers, and so on. The clarification of such conflicting definitions of basic terms is important because it might help us circumvent foreseeable misunderstandings in future political regulations.

For the understanding of algorithmic cultures, it is important to understand the multiplicity and entanglement of these imaginaries, epistemic views, practical usages, and performative consequences. For this reason, scholars in social sciences, cultural studies, and in particular, cultural sociology, should take heed and not mix up or conflate promises, imaginaries, and practical effects. This is not to say that we are reducing imaginaries to mere fantasies. Imaginaries are also real; they have real effects in algorithmic cultures, and thus need to be taken into account. However, the performative effects of imaginaries, and the performative effects of practices, do differ. It is important to be able to distinguish the two, and not only for cultural sociology.

References

Alexander, J. 1990. "Analytic Debates: Understanding the Relative Autonomy of Culture." In *Culture and Society: Contemporary Debates*. Edited by J. Alexander and S. Seidman, 1–27. Cambridge, UK: Cambridge University Press.

Alexander, J. 2004. "Cultural Pragmatics: Social Performance Between Ritual and Strategy." *Sociological Theory* 22 (4): 527–573.

Alexander, J. and Smith, P. 1998. "Sociologie culturelle ou sociologie de la culture? Un programme fort pour donner à la sociologie son second souffle." *Sociologie et sociétés* 30 (1): 107–116.

Alexander, J. and Smith, P. 2002. "The Strong Program in Cultural Theory: Element of a Structural Hermeneutics." In *Handbook of Sociological Theory*. Edited by J. Turner, 135–150. New York: Kluwer Academic/Plenum Publishers.

Amoore, L. and Piotukh, V., eds. 2016. *Algorithmic Life: Calculative Devices in the Age of Big Data*. London: Routledge.

Barocas, S. and Nissenbaum, H. 2014. "Big Data's End Run Around Consent and Anonymity." In *Privacy, Big Data and the Public Good*. Edited by J. Lane, V. Stodden, S. Bender, and H. Nissenbaum, 44–75. Cambridge, UK: Cambridge University Press.

Barocas, S., Sophie, H., and Ziewitz, M. 2013. "Governing Algorithms: A Provocation Piece." Presented at *Governing Algorithms: A Conference on Computation, Automation, and Control*, New York University, May 16–17.

Becker, K. 2009. "The Power of Classification: Culture, Context, Command, Control, Communications, Computing." In *Deep Search: The Politics of Search Engines Beyond Google*. Edited by K. Becker and F. Stalder, 163–172. Vienna: Studien Verlag.

Beer, D. 2013. *Popular Culture and New Media*. Basingstoke, UK: Palgrave Macmillan.

Beunza, D. and Millo, Y. 2015. "Blended Automation: Integrating Algorithms on the Floor of the New York Stock Exchange." SRC Discussion Paper, No 38. Systemic Risk Centre, The London School of Economics and Political Science, London.

Bolin, G. and Schwartz, A. 2015. "Heuristic of the Algorithms: Big Data, User Interpretation and Institutional Translation." *Big Data and Society* (July–December), 1–12.

Burrell, J. 2016. "How the Machine 'Thinks': Understanding Opacity in Machine Learning Algorithms." In *Big Data and Society* (January–June), 1–12.

Cardon, D. 2013. "Dans l'esprit du PageRank." *Réseaux* 1: 63–95.

CFTC 2015a. *Criminal Complaint, United States of America vs. Navinder Singh Sarao, AO 91(Rev.11/11)*.

CFTC 2015b. *United States of America vs. Nav Sarao Futures Limited PLC and Navinder Singh Sarao, Appendix to Plaintiff's motion for statutory restraining order containing declarations and exhibits*, Case: 1:15-cv-03398.

CFTC and SEC 2010a. *Preliminary Findings Regarding the Market Events of May 6th, 2010, Report of the staffs of the CFTC and SEC to the Joint Advisory Committee on Emerging Regulatory Issues*. 18. May 2010.

CFTC and SEC 2010b. *Findings Regarding the Market Events of May 6th, 2010, Report of the staffs of the CFTC and SEC to the Joint Advisory Committee on Emerging Regulatory Issues*. 30. September 2010.

Cheney-Lippold, J. 2011. "A New Algorithmic Identity: Soft Biopolitics and the Modulation of Control." *Theory, Culture & Society* 28 (6): 164–181.

Clarke, A. C. 1959. *Voice Across the Sea*. New York: Harper & Row.

Cliff, D. and Nothrop, L. 2011. "The Global Financial Markets: An Ultra-Large-Scale Systems Perspective." *The Future of Computer Trading in Financial Markets*. Foresight driver review, DR4. London: Foresight.

Cliff, D., Brown D., and Treleaven, P. 2011. "Technology Trends in the Financial Markets: A 2020 Vision." *The Future of Computer Trading in Financial Markets*. Foresight driver review, DR3. London: Foresight.

Daston, L. J. 2004. "Whither Critical Inquiry?" *Critical Inquiry* 30 (2): 361–364.

Davis, D. 2015. "@Amazon's Algorithms Are So Advanced, I've Been Offered Over 10,000 #PrimeDay Deals and Am Not Interested Any of Them" [Twitter Post], July 15, retrieved from https://twitter.com/DJD/status/621363180116267012 (accessed May 24, 2016).

De Certeau, M. 1974. *La culture au pluriel*. Paris: Seuil.

Deleuze, G. 1994. *Difference and Repetition*. Translated by Paul Patton. New York: Columbia University Press.

Eisen, M. 2011. "Amazon's $23,698,655.93 Book about Flies." April 22, retrieved from www.michaeleisen.org/blog/?p=358 (accessed May 24, 2016).

Ensmenger, N. 2012. "Is Chess the Drosophila Artificial Intelligence? A Social History of an Algorithm." *Social Studies of Science* 42 (1): 5–30.

Foresight 2012. *The Future of Computer Trading in Financial Markets: An International Perspective*. Final Report, Foresight, London.

Galloway, A. R. 2006. *Gaming: Essays on Algorithmic Culture*. Minneapolis, MN: University of Minnesota Press.

Galloway, A. R. 2012. *The Interface Effect*. Cambridge, UK: Polity Press.

Galloway, A. 2013. "Emergent Media Technologies, Speculation, Expectation, and Human/Nonhuman Relations." *Journal of Broadcasting & Electronic Media* 57 (1): 53–65.

Geertz, C. 1973. *The Interpretation of Cultures*. New York: Basic Books.

Gillespie, T. 2014. "The Relevance of Algorithms." In *Media Technologies: Essays on Communication, Materiality, and Society*. Edited by T. Gillespie, P. J Boczkowski, and K. Foot, 167–193. Cambridge, MA: MIT Press.

Goffey, A. 2008. "Algorithm." In *Software Studies: A Lexicon*. Edited by M. Fuller. Cambridge, MA: MIT Press.

Hallinan, B. and Striphas, T. 2014. "Recommended for You: The Netflix Prize and the Production of Algorithmic Culture." *New Media & Society* 18 (1): 117–137.

Hargittai, E. 2000. "Open Portals or Closed Gates? Channeling Content on the World Wide Web." In *Poetics* 27 (4): 233–253.

Heinich, N. 2009. *Le bêtisier du sociologue*. Paris: Klincksieck.

Hillis, K., Petit, M., and Jarrett, K. 2013. *Google and the Culture of Search*. New York: Routledge.

Honan, M. 2013. "I, Glasshole: My Year with Google Glass." *Wired*, December 30, retrieved from www.wired.com/gadgetlab/2013/12/glasshole (accessed May 24, 2016).

Introna, L. D. 2011. "The Enframing of Code: Agency, Originality and the Plagiarist." *Theory, Culture & Society* 28: 113–141.

Introna, L. D. 2016. "Algorithms, Governance, and Governmentality: On Governing Academic Writing." *Science, Technology, & Human Values* 41 (1): 17–49.

Introna, L. D. and Hayes, N. 2011. "On Sociomaterial Imbrications: What Plagiarism Detection Systems Reveal and Why It Matters." *Information and Organization* 21: 107–122.

Johnson, C., Dowd, T. J., and Ridgeway, C. L. 2006. "Legitimacy as a Social Process." *American Review of Sociology* 32 (1): 53–78.

Johnson, N., Zhao, G., Hunsader, E., Meng, J., Ravindar, A., Carran, S., and Tivnan, B. 2012. "Financial Black Swans Driven by Ultrafast Machine Ecology." Working paper, retrieved from arxiv.org/abs/1202.1448 (accessed May 24, 2016).

Kinsley, S. 2010. "Representing 'Things to Come': Feeling the Visions of Future Technologies." *Environment and Planning A* 42 (11): 2771–2790.

Kitchin, R. 2014. "Thinking Critically about and Researching Algorithms." In *The Programmable City Working Paper*, Maynooth, Republic of Ireland: Maynooth University, retrieved from http://papers.ssrn.com/sol3/papers.cfm?abstract_id=2515786 (accessed May 24, 2016).

Knorr Cetina, K. 2013. "Presentation to Panel, Theorizing Numbers." Presented at the *American Sociological Association Annual Meeting*, New York.

Kowalski, R. 1979. "Algorithm = Logic + Control." *Communications of the ACM* 22 (7): 424–436.

Kushner, S. 2013. "The Freelance Translation Machine: Algorithmic Culture and the Invisible Industry." *New Media & Society*. Published online before print January 3, doi: 10.1177/1461444812469597.

Lash, S. 2007. "Power after Hegemony: Cultural Studies in Mutation?" *Theory, Culture & Society* 24 (3): 55–78.

Latour, B. 1986. "Visualisation and Cognition: Drawing Things Together." In *Knowledge and Society: Studies in the Sociology of Culture Past and Present*, Volume 6. Edited by H. Kuklick, 1–40. Greenwich, CT: Jai Press.

Levy, M. 2011. *In the Plex: How Google Thinks, Works, and Shapes Our Lives*. New York: Martin and Schuster.

Mackenzie, A. 2005. "The Performativity of Code: Software and Cultures of Circulation." *Theory, Culture & Society* 22 (1): 71–92.

MacKenzie, D. 2006. *An Engine, Not a Camera: How Financial Models Shape the Markets*. Cambridge MA: MIT Press.

MacKenzie, D. 2015. "How Algorithms Interact: Goffman's 'Interaction Order' in Automated Trading." Working paper.

Mager, A. 2012. "Algorithmic Ideology: How Capitalist Society Shapes Search Engines." *Information, Communication & Society* 15 (5): 769–787.

Miller, C. C. 2013. "Privacy Officials Press Google on Its Glasses." *New York Times*, June 19, retrieved from http://bits.blogs.nytimes.com/2013/06/19/privacy-officials-worldwide-press-google-about-glass (accessed May 24, 2016).

Morris, J. W. 2015. "Curation by Code: Informediairies and the Data Mining of Taste." *European Journal of Cultural Studies* 18 (4–5): 446–463.

Nunes, M. 2011. "Error, Noise, and Potential: The Outside of Purpose." In *Error: Glitch, Noise, and Jam in New Media Cultures*. Edited by Mark Nunes, 3–23. New Haven, CT and London: Continuum.

Pasquale, F. 2015. *The Black Box Society: The Secret Algorithms That Control Money and Information*. Cambridge, MA and London: Harvard University Press.

Pirrong, C. 2015. "A Matter of Magnitudes: Making Matterhorn Out of a Molehill." In *Streetwise Professor* (blog by University of Houston finance professor Craig Pirrong), January 1, retrieved from http://streetwiseprofessor.com/?p=9337 (accessed May 24, 2016).

Pogue, D. 2013. "Why Google Glass Is Creepy." *Scientific American*, May 21, retrieved from www.scientificamerican.com/article.cfm?id=why-google-glass-is-creepy (accessed May 24, 2016).

Reckwitz, A. 2002. "Toward a Theory of Social Practices. A Development in Culturalist Theorizing", In *European Journal of Social Theory*, 5 (2): 245–265.

Reichertz, J. 2013. "Algorithmen als autonome Akteure?" In *SozBlog*, February 24, retrieved from http://soziologie.de/blog/2013/02/algorithmen-als-autonome-akteure/#more-964 (accessed May 24, 2016).

Roberge, J. and Melançon, L. Forthcoming. "Being the King Kong of Algorithmic Culture Is a Tough Job After All: The Justificatory Regimes of Google and the Meaning of Glass." *Convergence: The International Journal of Research into New Media Technologies*. Published online before print July 2, 2015, doi: 10.1177/1354856515592506.

Röhle, T. 2009. "Dissecting the Gatekeepers: Relational Perspectives on the Power of Search Engines." In *Deep Search: The Politics of Search Engines beyond Google*. Edited by K. Becker and F. Stalder, 117–132. Vienna: Studien Verlag.

Ruhe, N. 2014. "Algorithmic Cultures – Conference Report." *H-Soz-Kult*, October 29, retrieved from www.hsozkult.de/conferencereport/id/tagungsberichte-5626 (accessed May 24, 2016).

Ruppert, E., Law, J., and Savage, M. 2013. "Reassembling Social Science Methods: The Challenge of Digital Devices." *Theory, Culture & Society* 30 (4): 22–46.

Sandvig, C. 2015. "Seeing the Sort: The Aesthetic and Industrial Defense of 'The Algorithm'." *Journal of the New Media Caucus* 10 (1), retrieved from http://median.

newmediacaucus.org/art-infrastructures-information/seeing-the-sort-the-aesthetic-and-industrial-defense-of-the-algorithm/ (accessed May 24, 2016).

Sanz, E. and Stančík, J. 2013. "Your Search – 'Ontological Security' – Matched 111,000 Documents: An Empirical Substantiation of the Cultural Dimension of Online Search." *New Media & Society*. Published online before print April 29, 2013, doi: 10.1177/1461444813481198.

Savage, S. 2009. "Against Epochalism: An Analysis of Conceptions of Change in British Sociology." *Cultural Sociology* 3 (2): 217–238.

Seaver, N. 2014. "Knowing Algorithms." Presented at *Media in Translation 8*, Cambridge, MA, April 2013.

SEC 2013. Securities Exchange Act of 1934, Release No. 70694, October 16, 2013, Administrative Proceeding File No. 3–15570.

Seyfert, R. 2012. "Beyond Personal Feelings and Collective Emotions: A Theory of Social Affect." *Theory, Culture & Society* 29 (6): 27–46.

Seyfert, R. Forthcoming. "Bugs, Predations or Manipulations? Incompatible Epistemic Regimes of High-Frequency Trading." *Economy & Society*.

Striphas, T. 2009. *The Late Age of Print: Everyday Book Culture from Consumerism to Control*. New York: Columbia University Press.

Striphas, T. 2015. "Algorithmic Culture." *European Journal of Cultural Studies* 18 (4–5): 395–412.

The Social Media Collective. 2015. "Critical Algorithm Studies: A Reading List." retrieved from http://socialmediacollective.org/reading-lists/critical-algorithm-studies/ (accessed February 29, 2016).

Totaro, P. and Ninno, D. 2014. "The Concept of Algorithm as an Interpretative Key of Modern Rationality." *Theory, Culture & Society* 31 (4): 29–49.

Uricchio, W. 2011. "The Algorithmic Turn: Photosynth, Augmented Reality and the Changing Implications of the Image." *Visual Studies* 26 (1): 25–35.

Wansleben, L. 2012. "Heterarchien, Codes und Kalküle. Beitrag zu einer Soziologie des *algo trading*." *Soziale Systeme* 18 (1–2): 225–259.

Wasik, B. 2013. "Why Wearable Tech Will Be as Big as the Smartphone." *Wired*, December 17, retrieved from www.wired.com/2013/12/wearable-computers/ (accessed May 24, 2016).

Weiser, M. 1991. "The Computer for the Twenty-First Century." *Scientific American*, September 1, 94–100.

Zietwitz, M. 2016. "Governing Algorithms: Myth, Mess, and Methods." *Science, Technology & Human Values* 4 (1): 3–16.

2 The algorithmic choreography of the impressionable subject

Lucas D. Introna

Introduction

Advertising has become the dominant business model of the internet. As one of the early pioneers, Zuckerman (2014), suggests, it is "the entire economic foundation of our industry, because it was the easiest model for a web start-up to implement, and the easiest to market to investors." The fact that the business model of the internet is advertising is historically contingent. It is neither inevitable, nor the only possibility. Nonetheless, once the advertising business model became established as the default model, a certain logic flows from it. One might say that such a logic requires, or necessitates, that all the actors (in the world of the internet) become positioned in particular ways—be they users, technology developers, advertisers, digital entrepreneurs, etc. Specifically for us, advertising needs an audience. Not just any old group of people, rather, the right person, at the right time, to see the right advertisement. That is, it requires *impressionable subjects*. Subjects that are so impressed—pressed into or imprinted on—that they are highly likely to *convert*. That is, do something of value for the company whose advertisement it is—such as click on the advertisement, register on the site, buy a product/service, and so forth. Thus, in the business model of advertising, the users of the internet need to become produced or positioned as impressionable subjects, specifically—and such positioning requires a particular regime of knowledge (or truth), as Foucault (1991) would suggest. The human subject is not an impressionable subject from the start, as it were—impressionability is neither necessary nor originally founded. Such subjects need to be produced, or perhaps, more precisely, enacted (Foucault 1980). To produce these impressionable subjects, the ones that will convert, a complex choreography is needed—a choreography in which algorithmic agency is playing an increasingly sophisticated part, as we hope to show below.

For Foucault there is an intimate connection between power, knowledge, and subjectivity (Hall 1997). He suggests that power is relationally enacted and productive. Power is not an origin, but rather the outcome of the ongoing relational positioning of subject/objects within *material discursive practice* (Barad 2007; Foucault 1978, 94). Such positioning becomes constituted through regimes of knowledge. Knowledge is understood here as that which can be *produced* through a series of methods, techniques, and technologies. These methods, etc., include mechanisms for inscription, recording, and calculation—that is, diverse ways of observing, and of encoding, subject/object positions. Through these domains of knowledge, subjects can become amenable to intervention and regulation—they can become positioned, or governed, in specific ways (Foucault 1991). Thus, Foucault (1980, 52) concludes that "the exercise of power perpetually creates knowledge and, conversely, knowledge constantly induces effects of power. [...] It is not possible for power to be exercised without knowledge, it is impossible for knowledge not to engender power." Moreover, power "produces reality; it produces domains of objects and rituals of truth. The individual and the knowledge that may be gained of him belong to this production" (Foucault 1977, 194). In this ongoing circulation of power and knowledge the subject becomes enacted or positioned, and governed, in particular ways, in particular material discursive practices. For example, using specific methods and techniques—such as IQ tests, progress tests, classroom observations, etc.—some students become positioned as 'good' students, and others as 'struggling' students, in the material discursive practice of education. Over time, such positioning becomes the taken-for-granted—one might say, the valid and necessary—material discursive frame, relative to which subjects negotiate their own positioning, or position themselves. That is to say, as the way they take themselves to be, within such a regime of knowledge—'I am a good student' or 'I am a struggling student.'

In our discussions below, we want to show how algorithmic actors emerge as producers of particular domains of knowledge, using very specific—and historically contingent—mechanisms of inscription, recording, and calculation, which position internet users in specific ways, in order to enact them as particular impressionable subjects. Specifically, as algorithms produce knowledge of us (indirectly through our online behavior as journeys) we become positioned—also by ourselves—as this or that type of subject—for example, one that is active, likes sport, listens to particular music, etc. Indeed, what makes online advertising different to other media is the diversity of methods, techniques, and technologies (mostly algorithmic) for the production of a particular domain of knowledge—that in turn function to choreograph certain subject positions, meticulously. Based on this knowledge, we are shown advertisements, or not, by the algorithms. Through these advertisements, we also get to 'know,' and position, ourselves. Hence, over time, as we become positioned, and start to position ourselves, in particular ways—we find ourselves, and taken by other, to be subjects that need, want, or desire those products shown to us in these advertisements. It is of course not the case that these algorithmic actors *make* us become these subjects, it is rather that the regimes of knowledge—based on historically contingent mechanisms of inscription, recording, and calculation—produce the very conditions under which our subjectivity becomes negotiated, and, freely taken up by us, as being this or that type of person. Thus, rather than taking the subject as an individual with some reducible and internal core of meaning (beliefs, needs, desires, etc.), Foucault's work on power/knowledge suggests that the subject is produced historically and contingently—in and through regimes of knowledge. That is, the subject is constituted through being positioned in correlative elements of power and knowledge (algorithmically produced knowledge, in our case). In such positioning: "[c]ertain bodies, certain gestures, certain discourses, certain desires come to be constituted as individuals. The individual ... is I believe one of [power's] prime effects" (Foucault 1980, 98).

In considering the production of the impressionable subject, in online display advertising, we will be interested in the production of power/knowledge through the flow and circulation of agency in and through the *sociomaterial whole, of the internet*. Agency does not just flow through humans, it also flows through non-humans, as suggested by Latour (1988, 2005)—specifically, in our case, algorithms. In the sociomaterial whole of the internet agency is always borrowed and translated from elsewhere; and is never at the singular bidding of any human or non-human actor per se (Latour 2005, 46). The impressionable subject is produced but there is no producer, as such. Thus, tracing the flow of agency through a heterogeneous array of actors in the sociomaterial assemblage (or the 'agencement' as Çalışkan and Callon [2010] would call it) of the internet is difficult, if not impossible—even more so for digital actors (Introna 2006; Introna and Nissenbaum 2000; Introna and Wood 2004). Digital agency is often subsumed, inscrutable, and opaque, even if one can read the code (Barocas *et al.* 2013). Still, tracing the flow of agency through socio-digital assemblages (including algorithms) is important because these actors do *not only act they also simultaneously enact*—to be more precise, they are *performative* (Barad 2007; Pickering 1995; Butler 1990). In other words, it does not just trace the subject, it produces the subject, as was suggested above. Or, as Whitehead (1978, 23) suggested: "[H]ow an actual entity becomes constitutes what that actual entity is. [...] Its 'being' is constituted by its 'becoming.' This is the principle of process." Thus, in tracing the algorithmic choreography of the impressionable subject, below, we will attend to the 'technical details,' though not as *mere* technical details, but as performative material discursive practices (Orlikowski and Scott 2015). That is, as historical and contingent mechanisms of power/knowledge that progressively enact the impressionable subject in increasingly complex, and very specific ways. In short: as circuits in which power, knowledge, and impressionable subjects co-constitutively circulate—in and through the flow and circulation of agency (Ingold 2011).

The flow of the narrative of the enactment of the impressionable subject, presented below, is in a sense historical—but not as a linear story that somehow adds up to the history of online display advertising. We rather highlight what seems to us to be important constitutive enactments, which can shed light on the ongoing production of the impressionable subject. In a way, we attempt to do a sort of *genealogy* in the Foucauldian sense (Foucault 1984). That is, we take careful note of the seemingly insignificant and historically contingent 'technical' practices, which are in fact constitutive of the becoming of the impressionable subject. This is not the only narrative possible, of course, and by no means the authoritative narrative. In articulating this narrative, we will focus on, or locate, four significant enacting moments that seem to us to be constitutive of the impressionable subject. We have named these enactments as, the production of: (1) *the gazing subject*, (2) *the animated subject*, (3) *the individuated subject*, and finally, (4) *the branded subject*. In what follows we want to show, in as exact detail as possible, how these subject positions are produced in the becoming of the sociomaterial whole, of the internet, in order to produce the impressionable subject.

Becoming the *gazing subject*: capturing the gaze through screening it

What would make a computer *screen* appear as a meaningful location for advertising? It is not obviously such. It seems clear that the early computer users—especially computer programmers—were not expecting advertisements to pop up on their computer displays. Perhaps this was because the computer display, for the programmer, was not a *screen*. It was rather a work surface—a place where code was created, and where algorithms were turned into outputs—mostly data operations or reports on printers, and so forth. Nevertheless, such a positioning—of the computer display as a 'screen'—might have become more apparent as computers shifted from larger mainframe computers to personal computers. With this shift, the screen became repositioned from a place where you created code, and the like, to a place where you received relevant outputs, results, etc. With this repositioning the work surface transforms into a 'screen' (Introna and Ilharco 2006). Screens, as screens, can grab and hold our gaze. This is because screens 'screen' (in the sense of filtering and organizing) what is supposedly relevant, to those assumed to be in front of it, within a situated context. A screen frames and enacts a distinction (a boundary) between what is relevant and what is not, in a particular situated context. Think of the screens in airports, lecture rooms, train stations, and so forth. Positioned as screens these surfaces 'screen,' that is, automatically prioritize and locate our attention—in the specific situated context. For advertising to work the gaze of the subject needs to be produced as disciplined, i.e., located in a reliable manner. Or, more accurately, the subject needs to be produced as a gazing subject. Enacting the computer screen as a 'screen,' as a place where what is supposedly relevant appears, is the first step in doing that. This locating of the gaze of the subject, in a more or less reliable manner, transforms that gaze into a locatable object—which is, according to Callon and Muniesa (2005), the first step towards becoming a market as well. Locating the gaze is the first step in making the gaze *calculable*. It is worth noting that the pervasiveness of computing screens (computers, tablets, smartphones, etc.) that locate and hold our gaze in contemporary society has had a dramatic impact on the shift of advertising from traditional media to digital screens (Evans 2009). How is this enactment of the gazing subject achieved, algorithmically?

The first systematic online display advertisements appeared in 1990. This was made possible by *Prodigy*. Prodigy was a joint venture between IBM, a technology company, and Sears, a retail company—this is significant to note. The subject imagined in this joint venture was an impressionable subject from the start. Prodigy created a sort of internet service that was delivered through a standard custom designed *graphical user interface* (GUI)[1] (see Figure 2.1). This meant that through the GUI access to a whole host of content could be carefully curated, by the developers and advertisers—this was prior to the development of HTML and HTTP by Berners-Lee. This carefully curated content established a potentially reliable relation between content and the gaze of the subject—as long as that gaze could be held, of course. It is interesting to note the similarity between the GUI design and that of traditional media such as magazines and newspapers.

With such a standard GUI a certain obligatory passage point (Callon 1986) was established. The delivery of a variety of services, such as e-mail, bulletin boards, news, etc., through this GUI screen, suggested relevance and interest, and positioned the user in front of it as a gazing subject. As such, Prodigy included advertisements—for this assumed subject—on the bottom fifth of their screen, which they sold to advertisers. It often consisted of a simple text message (with some very basic graphics) inserted at the bottom of the screen (see Figure 2.1). Every time a user logged on (or requested a page of content) the advertisement was also displayed. It was assumed that the top four-fifths of the screen would have content selected by the user, thus be of interest, and as such secure the ongoing gaze. The bottom fifth would then simultaneously deliver an

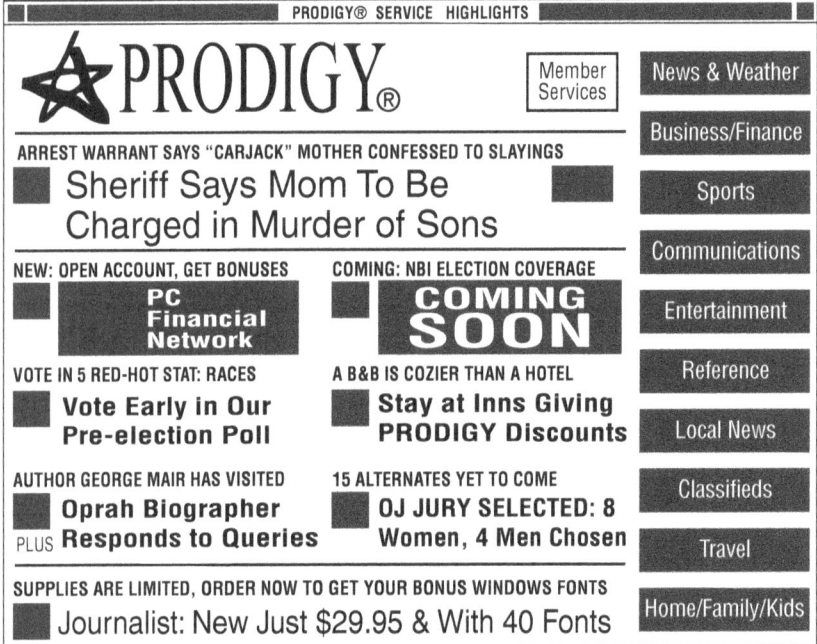

Figure 2.1 The Prodigy graphical user interface (GUI) with the bottom fifth advertising.

advertisement. It was possible for Prodigy to tell the advertisers how many times the advertisement, called an *impression*, was displayed. An 'impression' is a single viewing of a single advertisement by a single individual. Displayed impressions produced a calculable property that allowed for the enacted gaze of the subject to be sold as a cost per 1,000 impressions (CPM)—a market needs a currency and CPM was it, at least initially.

It is worth noting that from early on the terminology 'impression' was used. This suggested how the advertisers saw the agency and positioning of the subject in the sociomaterial whole. The purpose of the advertisement is to impress (make and imprint on) the subject, or make the subject an impressionable subject. This attempt to fix the gaze of the subject did not go unnoticed or unchallenged. Apparently, some users of Prodigy used to stick a piece of paper or plastic over the bottom fifth of the screen—this was possible because it was in a fixed location. The predictability or reliability of the gaze played both ways. One can imagine that they did not want to pay for a service that also aimed to enact them as an impressionable subject; this was not the deal, as it were. However, it seems that there was already an implicit 'deal' and maybe they were unaware of it. A Prodigy executive suggested, in 1991, that 'Every time you use the service to buy a holiday gift, book an airline ticket, pay a bill, trade a stock, send flowers or buy stamps, you are helping to assure the continuation of a flat, unmetered fee.'[2] This was in response to the outrage when Prodigy started to charge if a user sent more than 30 e-mails per month (25c per additional e-mail). Thus, his suggestion was that you could get all these services (e-mail, bulletin boards, etc.) but only if you shop—that is, become an impressionable subject.

Other services, such as AOL, also offered standard GUIs that also allowed for the enactment of a gazing subject (thus, producing impressions that could be sold). The problem for advertisers was that these curated and choreographed locations only captured the audience *who were members of their service*, which was paid for. What was required was a standardized universal cross-platform GUI available to everybody—in short, a general, widely available 'shop' window for internet content (what we now call a World Wide Web *browser*). A universal and standardized window for content that can reliably enact the gazing subject, anywhere content is being accessed.

The Mosaic World Wide Web Browser Version 1.0 was released in September 1993 by the National Center for Supercomputer Applications (see Figure 2.2). The name already suggests the subject being enacted; it is a 'browser,' and what this window into the internet enables is 'browsing.' Mosaic had a number of significant innovations that were very important for it to become the preferred web browser. First, it was easy to install and was a *cross-platform or multi-platform browser*. Second, Mosaic made it possible for images and text to appear on the same page (through its <IMG> tag). Earlier browsers allowed images to be included but only as separate files (which could be opened on request). Third, it consisted of a GUI with clickable buttons which allowed users to navigate easily as well as controls that allowed users to scroll through text with ease. Finally, and important to our discussion, it had a new way of embedding

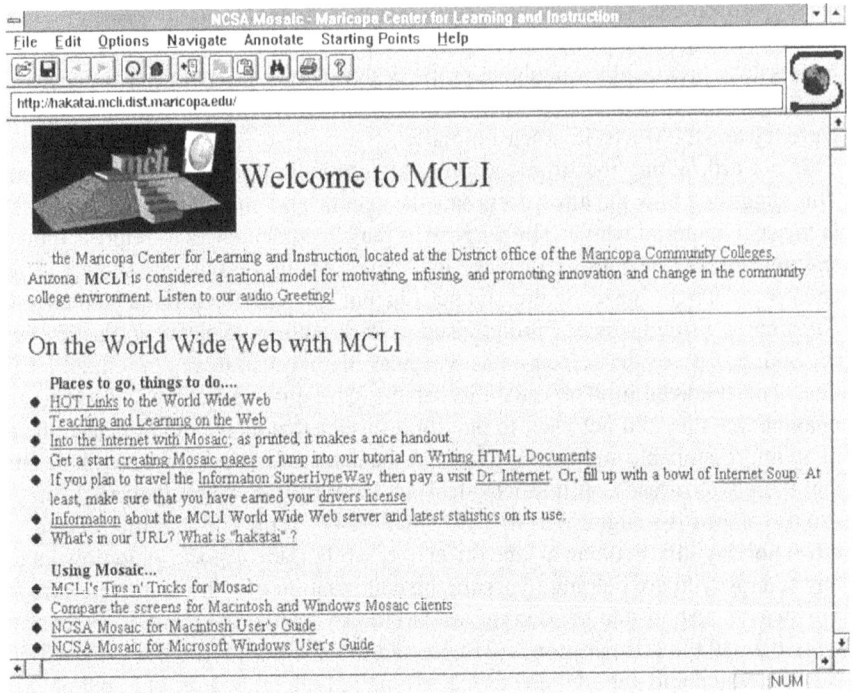

Figure 2.2 The Mosaic World Wide Web browser (in 1994).

hyperlinks (as highlighted and underlined text) which facilitated what we now refer to as 'clicking' on a hyperlink to retrieve content. In earlier browsers hypertext links had reference numbers that the user typed in to navigate to the linked document or content.[3] The new embedded hyperlinks allowed the user to simply click on an object (text or image) to retrieve a document. Mosaic was described by the *New York Times* as "an application program so different and so obviously useful that it can create a new industry from scratch."[4] With this standard, cross-platform browser, the user's experience of content can be curated dynamically to a significant degree.

More significantly, with this browser with embedded hyperlinks the user somehow becomes the 'curator' of the content on the screen. Through clicking on links subsequent content was determined by the user, allowing for individualized journeys through content, all curated through a standardized GUI (the web browser). This was a very significant development in the production of the impressionable subject. Traditional media, such as radio, television, and newspapers, are all 'push' media. In other words media content is determined by the publisher. It was easy therefore for the viewer to disregard content as 'irrelevant'—and as such lose the gaze of the subject. In the hyperlinked data structure (made possible by the development of HTML) content can be 'pulled' by the user (or browser) according to their own choices, or so it seems. Thus, this 'pull' model enacts a subject that seems to enact its own choices, curates its own content, according to its own interests. As such most of what appeared on the screen was seen as 'relevant,' also, perhaps the advertising. Of course, such choices are also subtly curated by the publisher through the links available on a particular page, etc. Hence, there is a subtle enactment of agency involved. Nevertheless, through hyperlinking, not only relevance but also, significantly, increasingly individualized subjects are produced, ones that can be categorized according to his/her browsing journeys, which is very important for the production of the impressionable subject, as we shall see. More specifically, a variety of *mechanisms and techniques for the production of knowledge* were being put in place.

Banner advertisements were popping up everywhere on the screens of users. But did they impress? Only advertisements enacted as 'impressions' would enact animated subjects, subjects that might convert. Maybe the sheer volume of advertisements popping up on their screens turned the gazing subject into a bored subject? A mechanism of knowledge was necessary to establish whether the gazing subject has indeed become an *impressed* subject.

Becoming the *animated subject*: producing 'interest' as clicking-through

The first clickable banner advertisement on a computer screen was launched on October 24, 1994. The advertisement was for AT&T and displayed on the Hotwire website (which was the precursor to *Wired* magazine). It simply said "Have you ever clicked your mouse right here" with an arrow to "you will" (see Figure 2.3).

Apparently 44 percent of those who saw the advertisement clicked on it, which was very significant but probably connected with its novelty value. Nevertheless, with the click-through action advertisers had their first currency (or metric) to claim that locating the gaze of the subject actually matters, that the subject was impressionable. With the clickable advertisement the gaze of the subject is transformed. Clicking on the advertisement produces the potentially 'interested subject.' The subject is now "endowed with a [new] property that produces distinctions" (Callon and Muniesa 2005, 1235).

Clicking produces an animated subject, and a feedback loop, which means advertising campaigns can now be monitored and optimized. For example, by

Figure 2.3 AT&T clickable advertisement shown on Hotwire (1994).

moving an advertisement around and seeing how many clicks it attracts in particular locations on the screen, etc. The early banner ads (such as the one in Figure 2.3) were hard-coded (or directly inserted) into the HTML code of the publisher's webpage. This approach was inflexible and did not allow for the dynamic serving of banner ads, or for the on-going collection of data about click-through rates, etc. For that an important new actor was needed in the socio-material assemblage, *the ad-server*. The ad-server will insert an advertisement dynamically into a predefined standardized location, or locations, on the page requested by the browser. Initially ad-servers were co-located with the publisher's web server. It was only a matter of time before this function was outsourced to external ad-servers (or third-party ad-servers, as they are often called). One of the first remote 'centralized advertising servers' was SmartBanner by a company called Focalink. Their press release on February 5, 1996 claimed that SmartBanner

> can hold thousands of different advertising messages and service hundreds of web sites simultaneously. It permits marketers to take control of their Internet advertising and to deliver the most appropriate message to selected audience segments. [...] Until now, marketing on the Internet has at best taken a shotgun approach.[5]

This rapidly developing environment (where ad content was rendered into browsers dynamically) also needed standards (of size and location of creative content) to facilitate its development. In 1996, a set of standards for display advertising was produced by the newly established Interactive Advertising Bureau (IAB) which allowed advertisements to be rendered seamlessly in web browsers (Goldfarb and Tucker 2014). This standardization was not always seen as good by the advertisers as it also facilitated the creation of potential standardized spaces of irrelevance, such as the 'gutters' on the sides of pages. This would be places on the screen where ads are expected and therefore seen as less relevant or not relevant at all (Goldfarb and Tucker 2014).

As click-through banner advertisements proliferated, together with the explosion of publishers—as every producer of a webpage was potentially a publisher—click-through rates started to fall.[6] For example, in 2011 a display advertisement would expect to gain around one click for every 1,000 impressions (0.1 percent).[7] Moreover there is plenty of evidence to suggest that clicking-through did not necessarily become a good proxy for the likelihood that the subject had become an impressionable subject (i.e., is likely to convert).[8] Finally, a significant number of actors realized that one could generate advertising income by generating click-through traffic using software programs called clickbots, software that enact what is known as click fraud. A whole new set of actors emerged to try to detect the difference between an impressionable subject and the unimpressed clickbot (Walgampaya and Kantardzic 2011).

Animating the subject is not enough to enact the impressionable subject. The animated subject needs to be transformed into a subject that can be positioned more specifically, that is the subject needs to become individuated. What was needed was more detailed knowledge—knowledge that would enable a more specific positioning of the subject.

Becoming an *individuated subject*: producing a shadow through cookies

To produce an impressionable subject—and not just a 'natural born clicker' (one that clicks for the fun of it)—requires more than fixing the gaze and animating the assumed subject. It requires, perhaps, a much more subtle choreography between a particular subject and a particular advertisement, in order to produce relevance, and interest, in some way. Perhaps by tracking it, or stalking it? The clicker needs to become individuated, or at least an instance in a particular set of categories. Such tracking and stalking actions would require a stream of *sustained and continuous interactions over time*, a significant challenge in a world based on stateless protocols. Both Internet Protocol (IP) (the foundation for the Internet) and the Hypertext Transfer Protocol (HTTP) (the foundation of data communication for the World Wide Web) are stateless protocols.

In 1994 Netscape released an internet browser[9] which incorporated a new technology called 'cookies.' The full name was 'Persistent Client State HTTP Cookies' (Kristol 2001). The practical problem they needed to solve was the lack of 'memory' in client/server interactions (as part of the e-commerce applications they were developing). Before cookie technology every recurrent interaction between a client and a server were as if it were the first time—like buying from a vending machine. Computer scientists refer to this as 'statelessness.' The interaction is 'stateless' in the sense that each request is treated completely independently of any previous one. This is an important design feature of the network, which provides it with significant flexibility and resilience. However, in statelessness the multiple interactions required to complete a multi-stage transaction could not be held together (or bundled together) as part of the same transaction—as for example in an e-commerce transaction. To solve this lack of interactional 'memory' over time a cookie (a sort of temporary identity tag) was placed on the client machine that identified it and stored some information that would enable the server to keep track of their prior interactions, if and when the client returns. For example, if a browser/user visits the *New York Times* site (server) it places a cookie on the computer that made the request (the client). Only the *New York Times* (*NYT*) can read or update its own cookie, no other actor could (this is called the same-origin policy). This seems reasonable since the browser/user initiated the interaction by visiting the *NYT* site and presumably would want their interaction to be smooth and effortless in the future, should the browser return.

However, with the use of remote third-party ad-servers the browser is also indirectly (and unknowingly) interacting with *third parties* (who are involved in inserting advertisements into the content that the *NYT* is serving it). In

Netscape's cookie technology these third parties (such as the ad-servers) are also able to place and read cookies on the browser's machine. Thus, every time the browser visits a site, which has advertising of one of the clients of the ad-server (or ad agency) on it, the ad-server can update their cookie record and as such develop a comprehensive picture of which sites the browser visits—that is to say it can 'observe' the browsing behavior of the internet browser, without the knowledge of the browser involved. One might say that the third-party ad-servers can 'stalk' the user as he or she browses. For example, in Figure 2.4 are listed 16 third-party cookies that were placed on a computer when it visited the *Guardian* newspaper's website (identified with an add-on called Ghostery). Through these 'tracks' the browser/user can become segmented and produced as an individuated subject who might be interested in a particular type of advertisement, rather than merely being a clicker.

It is worth noting that when standards were developed for cookies, by the Internet Engineering Task Force (IETF), this third-party cookie loophole, introduced by Netscape was retained despite its controversial nature (Shah and Kesan 2009). One reason for this was the fact that the Task Force took so long to get agreement on its proposed standards that an internet advertising business model had already emerged that had the cookie technology as its necessary condition.

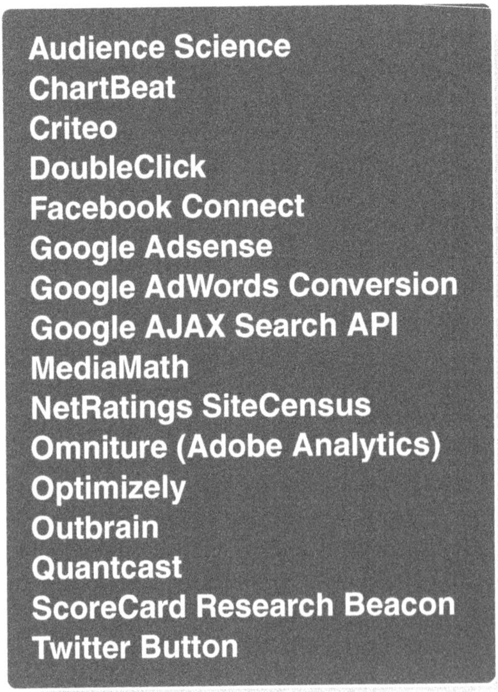

Audience Science
ChartBeat
Criteo
DoubleClick
Facebook Connect
Google Adsense
Google AdWords Conversion
Google AJAX Search API
MediaMath
NetRatings SiteCensus
Omniture (Adobe Analytics)
Optimizely
Outbrain
Quantcast
ScoreCard Research Beacon
Twitter Button

Figure 2.4 Third-party cookies when visiting the *Guardian* newspaper's website.

During this drawn out period the Netscape implementation of cookies had become the de facto standard and there was enormous pressure from the advertising industry to keep it that way. As such it became more or less adopted in the final proposals of the IETF, with some more or less significant modifications (Kristol 2001).

Combining cookie data with other sources of data (from traditional data aggregators) allowed ad-serving companies to do a fine-grained tracking and targeting of their advertisements—or so they suggested. The first company to develop a sophisticated tracking and targeting system using cookie technology was DoubleClick. They called it the 'Dynamic Advertising Reporting and Targeting' (or DART) system. This system automated a lot of the media buying and selling cycle and allowed advertisers to track their campaigns over a large number of publisher sites. It also became a catalyst for an approach that has become known as *behavioral targeting* (Jaworska and Sydow 2008) based on cookie technology.

Another approach, pioneered by Google, is to produce the impressionable subject through association with certain keywords (called AdWords). Advertisers create advertisements and assign keywords (bought on auction) thought to be relevant to their products and services (Chen *et al.* 2009). When these assigned keywords are entered by a user/browser the advertisement appears next to the search engine's generated results, allowing a browser to click on the ad in order to be redirected to the company's website. According to Google, this allows AdWords users to advertise "to an audience that's already interested in you." AdWords are also embedded in Google's other applications such as Gmail and YouTube. For example in Gmail they would scan the email for keywords and serve ads related to those keywords. This approach becomes known as *search and contextual advertising*.

The impressionable subject becomes individuated and produced through its hyperlink journeys and its search queries. *The impressionable subject is produced as what is being visited and what is being searched.* Significantly, this subject is mostly produced without the active involvement of the human. There are very significant algorithmic actors involved, a subject that is beyond the content we can cover here (Castelluccia 2012). There is also an increasing awareness of these stalking algorithms—and attempts to subvert them. For example, increasingly users/browsers are using algorithmic actors such as Trackmenot to obfuscate these stalkers (Nissenbaum and Brunton 2013). Opting out of tracking is also possible but often the cost of doing so is unclear and never straightforward. There are also significant efforts being made to move beyond the 'cookies' paradigm, such as browser fingerprinting (Nikiforakis *et al.* 2014).[10]

Becoming a *branded subject*: brand affinity and the meticulous choreography of agency

> Brand – German 'Brand,' also to burn. The verb sense 'mark with a hot iron' dates from late Middle English, giving rise to the noun sense 'a mark of ownership made by branding' (mid-17th century).
>
> (OED)

State data, or cookies, in a world of exponential increases in internet traffic, generate data, and a lot of it.[11] Also with dynamic ad-servers, every webpage can potentially be a place for publishing (what is referred to as *ad inventory*), irrespective of session depth.[12] This explosive proliferation of places to publish advertisements, or ad inventory, makes it difficult to discipline and hold the gaze of the subject, that is, to enact a reliable relation between the gaze of the subject and the advertisement such that the subject would be constituted as being impressionable. The complexity of choreographing and encounter—due to the proliferation of ad inventory, as well as the explosion of tracking data—requires new types of technology to produce a more intricate knowledge of the subject.

What is needed is a careful choreography of agency between the *individuated subject*, the *individuated content* and the *individuated place* (and at the right price). In this choreography of agency *timing is critical*. For this choreography three new (and interconnecting) actors are required. First, an actor to individuate the subject into a particular and very specific impressionable subject (learning algorithms), second, an actor that can buy/sell a *single impression (at the right place and price) in real time* (real-time bidding algorithms), and finally, an actor that can dynamically create creative content, for this particular subject at this very particular time. This way the already impressionable subject and the correlated creative (the ad) can be curated and choreographed to face each other exactly at the right time (in *real* time) and at the right price—in short: the meticulous curation of the impressionable moment which will *brand* the subject. How is this ontological choreography achieved exactly?

In discussing the enactment of this sociomaterial assemblage, we will focus on the practices of a particular company which used to be called Media6Degrees but is now called Dstillery. This is for two reasons: (1) the chief scientist of Dstillery, Claudia Perlich, has generously published a number of papers describing their approach (Perlich *et al.* 2014; Perlich *et al.* 2012; Dalessandro *et al.* 2012; Raeder *et al.* 2012); and (2) Dstillery does *prospecting* rather than retargeting. Prospecting is where advertisers target subjects with whom they have had *no prior interaction* (but who they think might be impressionable) in order to 'improve the penetration of their brand.' Most of the choreography we discussed above was based on *retargeting*—this is where the advertisers target subjects with whom they have interacted before (for example, they might have previously visited their website). In prospecting, the algorithmic production of the impressionable subject is particularly significant and indicative of where online display advertising might be heading.

Producing the individuated 'branded' subject

To transform the internet tracks (or 'brand-specific signals') of the browsing subject into that of an impressionable subject machine learning (or 'award-winning data science') is deployed. Machine learning requires lots of data, a very powerful computing infrastructure, and mathematical machine learning algorithms. Machine learning works inductively, as opposed to traditional artificial intelligence approaches, which are deductive. That is, it starts with actual behavior (or at least data generated or emerging from such behavior). Through statistical methods it tries to build models of such experience/behavior in order to, for example, predict future behavior, categorize or classify behavior, match similar behavior, cluster similar behavior, and so forth.[13] These models can then be subject to ongoing and automatic revision (or learning), based on new data of actual experiences/behaviors. The higher the quality and quantity of data they have to work from the better they perform. That is why Google wants us to interact with its applications 'for free,' because these interactions generate data, and data allows for a finer-grained individuation of the subject. The more individuated the subject the more valuable it is, in terms of impressionability. Through machine learning algorithms such individuation can happen automatically in order to produce the unique, highly impressionable subject, a market of one. According to Dstillery:

> A click. A tap. A swipe. A GPS query. They're not just actions, they're signals.... We collect data from the complete consumer journey—digital behaviors and real-world behaviors alike. We then extract the most important part: the patterns of shared behaviors that demonstrate a proven connection to your brand.... *We use these patterns to distill your purest possible audience*, the people who are scientifically proven likely to engage with your brand.[14]

Dstillery has partner agreements that cover approximately 100 million websites. Some of these sites are companies for whom they manage advertising campaigns, such as AT&T, Adobe, British Airways, Verizon, Best Western, American Express, etc. In 2012 they were working with 300 companies (or 'brands' as they refer to them). On these client sites they place their *action pixels* and on the data partner sites they place *mapping pixels*. Pixels (or pixel tags) are HTTP requests issued by a webpage to download invisible content in order to create an HTTP transaction, thus allowing for the setting of state data (or cookies). The pixel tag does not have any visual effect on the particular webpage. It is only there to provide Dstillery (as a third party) the opportunity to place their cookie on the browser's site (or update a cookie they had created earlier). Action pixels provide them with data on what is done on their client sites (that is conversion actions) and mapping pixels create data about the browser's 'journey' of websites visited—its browsing history. They use this data to create a linear regression model where the mapping data define the features (the x values) and action data define the class (y values or labels). These are massive models as they represent the 100 million possible data points in their observed data space (i.e., the 100 million URLs they monitor). They might look something like this:

$$APi = aMP1 + bMP2 + cMP3 \ldots + nMPN \text{ (where N} = 100 \text{ million)}$$

The basic underlying logic of their learning models is that the set of most recent sites visited (the journey) is a very good proxy for conversion (Dalessandro *et al.* 2012). They basically propose that if I have available a model of sites visited by a person (the features according to mapping pixels) who converted (the action pixels) then another person with a significantly similar set of sites visited will have an equally high probability of converting if shown the relevant ads (related to the initial conversion)—this allows them to locate "people on an accelerated path to conversion."[15] They claim that the *most recent* set of sites a browser visited 'captures' his/her current interests, needs, wants, and desires—even better than recent purchase behavior or long-term historical behavioral models, typically used in traditional marketing (Dalessandro *et al.* 2012). A browser that correlates very closely to the model will be tagged as having a high 'brand affinity.' Browsers will be ordered on a target list from highest brand affinity to lowest brand affinity. According to Perlich the subjects at the top of their target list (with a high brand affinity score) are "at least four times more likely to take [conversion] action than a random set of people who see that same ad."[16]

Thus, by building and using brand-specific reference (or trained) machine learning models a subject can be placed in a ranked list of potential targets (impressionable subjects) for every campaign run by the company. As suggested, at the top of the list would be the subject with a set of sites visited which matches most closely the reference model (or trained model) for that brand down to the one with the least overlap. In such a machine learning model each site will have a parameter to indicate its contribution to the predictive power of the model. Accordingly, not all site visits are equally important. The contribution of each site, as a data source for a particular model, is monitored on an ongoing basis to ensure that the site data adds predictive value to the specific brand model (Raeder *et al.* 2012). The ranked list of 'prospects' can be referred to as the target audience (or ranked list of increasingly impressionable subjects, for a particular campaign). Dstillery tracks approximately 100 million 'prospects' at any one time. The ranked lists for every campaign are refreshed on an ongoing basis as new data becomes available. Hence the model becomes increasingly more subtle in the distinctions that it makes. The initial model for every campaign is set using transfer learning of 'similar' products/campaigns (Perlich *et al.* 2014). It is important to note that these models work on the principle of correlation. That is, they assume we will tend to behave in a similar way (in terms of conversion actions) to others that behave similar to us in other dimensions (browsing journeys). This similarity constitutes us as having brand affinity or being impressionable.

Producing an individuated encounter (at the right price)

Once a particular browsing subject has been scored for all the brands they manage (ranked based on brand affinity) an opportunity to display a relevant advert must be found. Let us assume that this browser clicks on a link to retrieve a page from their often visited site such as the *NYT*'s (and there is potentially some advertisement space on the page). A call is sent to the ad-server, which passes this on to the ad-exchange (where buyer's algorithms and seller's algorithms meet). A call for bids is issued and a real-time auction is held. The winning bidder gets the right to display (in that particular session) a certain number of impressions at a certain price. When the content arrives in the GUI of the browser it will have the advertisement of the winning bidder inserted into it. This all happens in 100 milliseconds (the speed at which a human typically blinks is 300 milliseconds). The advertisement embedded might be a standard 'creative' designed for the campaign, or *dynamic creative optimization* can be used. This is where the creative is customized for the impressionable subject for a particular impression. Thus, instead of showing the subject the same advertisement, it is possible to show a series of advertisements that is customized to prime the subject before displaying a conditional offer, such as a 'for today only' advertisement.

The final stage in this agencement or assemblage is to monitor conversions and to update the brand affinity scores as well as the models—that is, to learn from this particular experience. For example, if a browser with a high brand affinity score does not engage in a conversion action, then the models would be adjusted to take that data into account. This updating and monitoring is all done automatically. Moreover, other machine learning models monitor the data streams to make sure that the integrity of the algorithmic whole is maintained (Raeder *et al.* 2012), for example, to eliminate fraudulent data from clickbots, to correct for sudden changes in data providers, and so forth. These three steps—brand affinity scoring, real-time bidding, and creative optimization—complete the meticulous choreography and curation of the impressionable subject to become exactly that: the right person at the right time with the right creative content (and at the right price). One might say a subject that has become fully and completely *branded*. It is important to note that this happens automatically (using machine learning algorithms) and that it is meticulously choreographed in a timeframe of milliseconds (between you clicking on a page and the curated content arriving in your browser). This requires an enormously complex technical infrastructure, which has been simplified here for the sake of brevity and the flow of the discussion. Enacting the impressionable subject is big business[17] and in many ways the cutting edge of technological innovation of the internet giants.

Some thoughts on the algorithmic choreography of the impressionable subject

[Y]our customer is ready to decide. All they need is a nudge.

(Dstillery website)

As was suggested above, the human subject is not an impressionable subject from the start. Such a subject needs to be enacted through a complex and subtle choreography involving a heterogeneous assemblage of diffused flows of agency. In the discussion above, we have tried to sketch the outline of such a choreography by highlighting a particular subset of flows that we might traditionally describe as 'technical' practices (or, more specifically, algorithmic practices). However, the claim is that they are never purely 'technical.' Rather, the claim is that these practices *already embody or impart the logic—or, one might say the intentionality—of the whole*, from the start, as it were. The discursive material whole (the internet) already embodies a particular logic (or business model) which enacts agency in particular ways. It is a logic or business model that is in fact based on advertising revenue. The business model or logic necessitates of advertising (and by implications the discursive material whole) to enact impressionable subjects, that is, ones that would convert. Consequently, from the very start technical innovations become imagined as sites or moments where this subject becomes enacted, as one of its necessary purposes. The creation of the graphical user interface (GUI) by Prodigy or the World Wide Web browser (Mosaic) is not just the creation of a useful interface, it is also immediately a mechanism to discipline the gaze, to fix and hold it, which is necessary to enact the subject as one that is attending to what is being displayed. This gaze, which flows in both directions, is not just about looking at something, as Foucault (2003)[18] reminds us. The gaze establishes relations of knowledge in which the subject becomes constituted, it is "no longer reductive, it is, rather, that which establishes the individual in his irreducible quality" (Foucault 2003, xv).

The hypertext (or hyperlink) structure allows for the subject to dynamically curate the content of what appears on that GUI screen as he or she transverses the vast content of the internet. This curation ensures *relevancy*. As such the screen holds the gaze since it 'screens,' that is, filters and organizes, what is supposedly relevant. These individually curated journeys create the knowledge that allows the algorithms to enact a subject as having 'interests,' 'needs,' and 'desires,' for example. The clickable image transforms the subject from a viewer to an animated or participatory subject—that is, a subject that is participating in the curation, not just of the content on the screen, but also of the advertising being displayed. Ad-servers (and especially third-party ad-servers) use click-through rates to optimize advertising placements. This curation of 'content' by the animated subject is not just a matter of what appears on the screen, it is also simultaneously the curation of the impressionable subject that the viewer is becoming. The development of 'cookie' technology solves the problem of stateless protocols, but it does much more besides. Third-party cookies placed by third-party ad-servers allow for the curated journeys (made possible by the hyperlink structure) to become individuated 'tracks' or traces of the 'who' that the 'I' that is surfing is becoming. The surfer is now not only curating content, they are also simultaneously curating a vast amount of knowledge about their supposed 'interests,' 'needs,' and 'desires.' Through animation, curation, and tracing an intimate set of *knowledge* of the supposed subject is being enacted. Through these domains of knowledge, the subject can become amenable to intervention and regulation—it can become positioned, or governed, in specific ways (Foucault 1991). As Foucault suggested, this knowledge/power "produces reality. [...] The individual and the knowledge that may be gained of him belong to this production" (Foucault 1977, 194).

Machine learning transforms all these tracks or traces into 'brand affinity.' Massive datasets of a 100 million dimensional matrices map the subject's internet journey onto dynamically evolving brand profiles to establish 'affinity.' In having a high brand affinity the subject is enacted as one that is branded to be 'like others'—who convert when shown branded advertisements. Dynamically customized advertisements are then delivered to the individuated or branded subject, bought in real-time auctions (between buyer's algorithms and seller's algorithms) lasting 100 milliseconds. The branded subject is a meticulously choreographed subject, enacted through a vast array of algorithmic actors, "which is four times more likely to convert," according to Dstillery,[19] than the unimpressed subject "shown a random advertisement." There is no doubt that the heterogeneous assemblage to enact the impressionable subject will evolve and become increasingly sophisticated. The logic of the whole—that is, the business model of the internet—will impart itself into every new innovation, it is a necessary condition of the becoming of the whole. Up to now we have followed this discussion from the point of view of the flow of algorithmic agency, let us now consider the flow of agency through the enacted subject more carefully.

Rose and Miller (2008, 139) in their discussion of the enactment of the 'consumer' in advertising practices (especially in relation to the work of the Tavistock Institute of Human Relations that they were discussing) suggest that what this work shows is that:

> In order for a relation to be formed between the individual and the product, a complex and hybrid assemblage had to be inaugurated, in which forces and flows imagined to issue from within the psyche of persons of particular ages, genders or social sectors were linked up to possibilities and promises that might be discerned within particular commodities, as they were organized within a little set of everyday routines and habits of life.

In the discussion above, this 'linking up' (of subject and product), in online display advertising, was demonstrated as the enactment of the impressionable subject through the choreography of a vast array of algorithmic actors. Yet, this choreography takes for granted reflective subjects. They are not simply dupes or mere fodder in the battle of corporate capital. Rather, they are reflexively aware of this choreography and actively involved in participating in it, more or less explicitly. In other words, they are also more or less actively trying to be the authors or curators of their own subject position, within the choreography—yet, never quite able to achieve it. As Latour (2005, 46) suggests, "action is borrowed, distributed, suggested, influenced, dominated, betrayed, translated. [...] [Sociomaterial agency] represents the major source of uncertainty about the origin of action."

The reasons for this 'major source of uncertainty' of agency (or diffused nature of agency) are manifold. For example, one might suggest that users or browsers can merely 'opt-out' of the tracking (or individuating) process—an option offered by most, including Google. They advise that:

> You can opt out of interest-based ads from Google via Ads Settings. When you opt out, you'll still see ads but they may not be related to factors such as your interests, previous visits to other websites or demographic details.[20]

Thus, if you opt-out you will see more or less random advertisements. Implied in this statement is the question: what would you prefer, meaningful, relevant ads based on your interests or random meaningless ads? The choice is yours. However, you cannot opt-out of advertising altogether. That deal—to have a 'free' internet funded by advertising revenue—was already made elsewhere, at some earlier stage by a whole host of agencies, directly or indirectly (Zuckerman 2014). What is clear is that the existence of the assumed 'free' internet is now dependent on the successful enactment of the impressionable subject, because that is the business model of the internet.

In this logic, implicitly agreed, it is assumed that it is in your interest to see meaningful ads (rather than random irrelevant ads). It is in Google's and other ad agencies' interest to make these ads meaningful (that is, to enact you as an impressionable subject). There is no option to pay for an advertisement-free internet (and who would you pay?)—and, indeed, it is suggested that most people will not be prepared to pay for such an advertising-free internet anyway (Curtis 2014). Thus, there seems to be a commonly agreed social contract that we are to become impressionable subjects, in exchange for a 'free' internet.

One might further suggest that this is not a problem as users can simply ignore the advertisements as they transverse the hyperlinks of the Web—i.e., they can also 'opt-out' of being enacted as an impressionable subject *by not attending* to these advertisements. In the Prodigy case, above, they could close the bottom fifth of the screen to avoid seeing the advertisements. It is also possible to use ad-blocker applications.[21] However, publishers suggest that "[v]iewing ads is part of the deal if users want content to be free" and that the use of these programs "breaks that implicit contract" (Mitchell 2014). Furthermore, in the contemporary assemblage the location and nature of the advertising content is dynamic and not necessarily clearly delineated from the other content. Moreover, ignoring it might not be a choice entirely available to the already enacted subject. There is a large literature of research in consumer behavior and psychology about what is known of the 'priming effect,' which might suggest some limits to the degree we are able to curate our enactment as subjects. This research suggests that prior (or concurrent) exposure to images and messages can enact a particular subject in a very significant manner—for example, the honest subject. A classic study in this regard is what has become known as the 'honesty box'[22] experiment (Bateson *et al.* 2006). It showed that images placed inconspicuously on a wall in a university staff kitchen (alternating between images of flowers and images of eyes) enacted subjects as more or less honest depending on the type of image, as shown in Figure 2.5. What is noteworthy is the way in which eyes (and different stares) seem to produce more or less 'honest' staff—that is, staff that would place money in the box—in ways that did not happen with flowers. Again, the suggestion here is not that the users are somehow duped or mere puppets on the strings of the advertisers. Rather the point is that agency, in subject positioning, is much more subtle and diffused than what we normally would tend to assume.

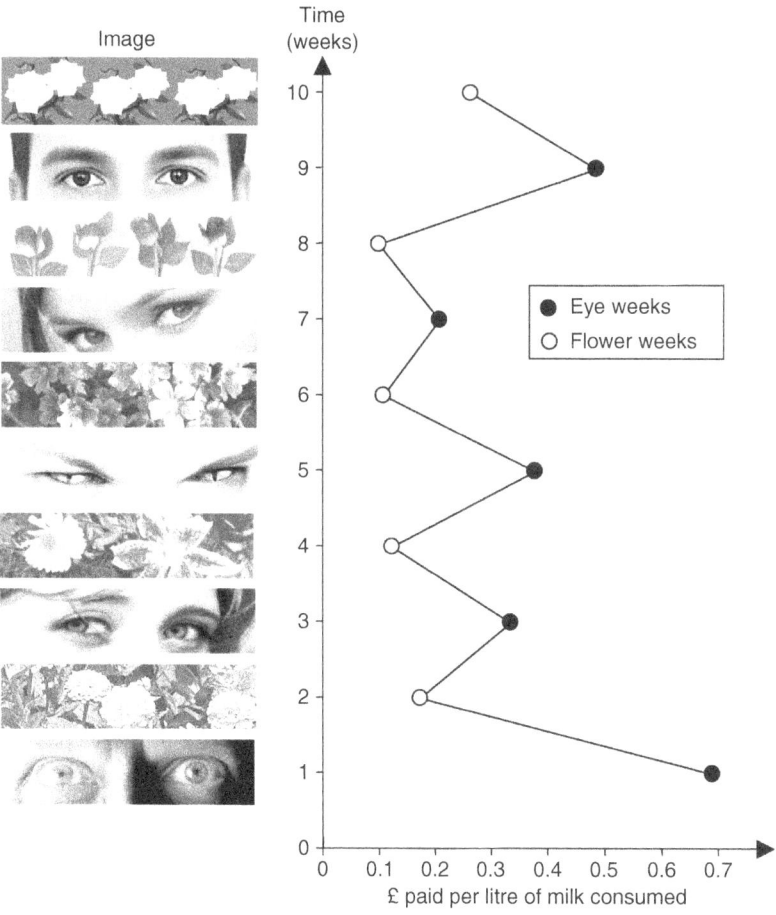

Figure 2.5 The honesty box experiment (taken from Bateson *et al.* 2006).

The impressionable subjects are not merely passive targets of advertising. They also often take an active role in the enactment of their own subjectivity—of course, within the discursive space offered by advertising, for example, in actively embracing the images and discourses of advertising—what Arvidsson (2005) calls 'informational capital'—to participate in the choreography and curation of their own subjectivity. The examples are endless: Apple products (iPhone, iPad, etc.) enact me as innovative; the Swatch watch enacts me as trendy, and so forth. The point is that subjects are very aware of the discourses of advertising and brands and their possibilities for enacting the latter as very particular (and perhaps desirable) subjects. It is what Foucault would describe as subjectivation:

> [s]ubjectivation is a process of internalization that involves taking a decision about being a particular type of subject. [...] It is an activity carried out by a human being as a subject who knows; for example, a subject who makes her or himself subject to truths circulating about [brands].
>
> (Skinner 2013, 908–909)

In this sense, the correspondence between subject and advertisement is not to be avoided but actively and explicitly embraced as informational capital at the subjects' disposal to form them into the subjects they want to become, if possible. As such the advertisements that appear on the screen are not 'mere' advertisements, they are also simultaneously suggestive of the subject that I want to become—that is, if the individuating assemblage enacting me 'got it right.' If, on the other hand, the advertisements appearing on my screen are not in line with whom I believe I am becoming then I might object, 'why am I shown this rubbish? What did I do to make these algorithms believe that I am such and such a type of person?' The point is that the enactment of the subject as an impressionable subject is also part of the enactment of the self as a particular subject through a careful curation of associations with brands—that is, an ongoing project of subjectivation: being both *subject of*, and *subject to*, the discursive material choreography.

What makes the flow and choreography of agency, in online display advertising, different from other discursive material practices is that power/knowledge is asymmetrically produced. The algorithms have intimate knowledge of the subject, yet the subject is quite ignorant of them. The subject is often located, branded, and served advertisements according to invisible categories constructed by machine algorithms. Most of the sociomaterial whole constituting it is a black box that is opaque even to the designers thereof (in the case of machine learning algorithms, for example) (Ziewitz 2016). Moreover, where do you go to complain or challenge it? Besides, even if you complain and challenge the designers, they might respond that it is merely machine learning algorithms doing it (the categories 'mean' nothing as such and are merely reflecting the data they are given)—i.e., they are algorithmically objective and neutral in their choices of categories (Gillespie 2014). This is what Google claimed when people complained about the 'racist' and 'bigoted' suggestions supplied by the auto-complete function (Baker and Potts 2013).[23] In a way, they might say that they (designers and algorithms) are also enacted in the becoming of the sociomaterial whole in which they are also non-original subjects being positioned in the ongoing choreography of agency. Clearly, the enactment of subject positions flow *in all directions simultaneously.*

To conclude: given that the dominant business model of the internet is advertising, the central argument of this chapter is that the performative enactment of the impressionable subject is a necessary condition of the ongoing becoming of the internet—it is the foundational logic of the sociomaterial whole. As such, this logic will impart itself into every innovation. The production of this subject will require increasingly detailed levels of algorithmically produced knowledge in order to position subjects as impressionable. One might say it will require an intimate and detailed choreography to continue to enact such positioning. This would suggest that the incentive to develop an increasingly intimate knowledge of the subject will only expand. The algorithmic choreography of the impressionable subject has only just begun—unless of course there is a sudden change in the business model of the internet, which seems unlikely. In our discussion above, we have traced the choreography of the flow and circulation of agency mostly through the flow of algorithmic agency. However, in the choreography of the impressionable subject there are many different agencies circulating—which overlap, coincide, correspond, etc. In these circulations intentions, identities, positions become translated and displaced in ways that do not allow any definitive choreographer to emerge (not the algorithms, nor the subject, nor the advertisers, nor the advertising agencies, and so forth). Furthermore, performativity flows in all directions, producing new flows and subject/object positions (such as ad-blockers, obfuscators, clickbots, and so forth). Most importantly, in this choreography the impressionable subject is no mere 'puppet.' Such enactments would not have worked if the subjects were not also already willing, and productive, participants in this performative choreography. What is hopefully clear from the above is that in this choreography agency is never a simple straightforward question, it is "borrowed, distributed, suggested, influenced, dominated, betrayed and translated [and] it represents the *major source of uncertainty about the origin of action,*" as Latour has suggested (2005, 46; emphasis added). This uncertainty means that there are no simple interventions available to those who want to regulate or govern such a choreography (such as, for example, more transparency, more privacy, etc.). To govern this choreography would require ongoing and detailed study of what the choreography is becoming—this account is a first attempt to do just that.

Notes

1 The development of the GUI interface in personal computing is an important precursor to this development. See Reimer (2005) for a genealogy of the GUI.
2 w2.eff.org/Net_culture/Virtual_community/prodigy_gaffin.article (accessed February 20, 2016).

3 The previous version ViolaWWW on which Mosaic was built also allowed for imbedded hypertext links but it was not a cross-platform browser (Windows, MAC OS, and Linux).

4 http://history-computer.com/Internet/Conquering/Mosaic.html (accessed February 20, 2016).

5 www.thefreelibrary.com/Major+Companies,+Ad+Agencies+Using+New+Internet+Ad+Server%3B+Focalink's...-a017920427 (accessed February 20, 2016).

6 www.comscore.com/Insights/Press-Releases/2009/10/comScore-and-Starcom-USA-Release-Updated-Natural-Born-Clickers-Study-Showing-50-Percent-Drop-in-Number-of-U.S.-Internet-Users-Who-Click-on-Display-Ads (accessed February 20, 2016).

7 www.smartinsights.com/internet-advertising/internet-advertising-analytics/display-advertising-clickthrough-rates/ (accessed February 20, 2016).

8 www.adexchanger.com/research/clickthrough-rate-rethink11/ (accessed February 20, 2016).

9 The Netscape browser was based on the Mosaic browser.

10 Test the uniqueness of your browser here https://panopticlick.eff.org/.

11 According to a Cisco report, annual global IP traffic will pass the zettabyte (1,000 exabytes) threshold by the end of 2016 (a zettabyte is equal to one sextillion (10^{21}) or, strictly, 2^{70} bytes). www.cisco.com/c/en/us/solutions/collateral/service-provider/visual-networking-index-vni/VNI_Hyperconnectivity_WP.pdf (accessed February 20, 2016).

12 Historically advertising space on the home page is considered more valuable and the deeper we go into a website (called session depth) the value of the ad inventory becomes less valuable.

13 Spam filters, recommender systems (Amazon, Netflix), Google autocomplete, etc. are all applications based on machine learning algorithms.

14 www.dstillery.com/how-we-do-it/ (accessed November 15, 2015), emphasis added.

15 www.fastcompany.com/1840817/media6degrees-knows-what-you-want-buy-even-you-do (accessed February 20, 2016).

16 www.fastcompany.com/1840817/media6degrees-knows-what-you-want-buy-even-you-do (accessed February 20, 2016).

17 Dstillery was identified by Forbes as one of America's most Promising Companies. Internet advertising revenue for 2013 was $42.8 billion for the USA alone. www.iab.net/about_the_iab/recent_press_releases/press_release_archive/press_release/pr-041014 (accessed February 20, 2016).

18 In *The Birth of the Clinic* Foucault talks about the gaze of the clinician, but it is equally true for the gaze of the subject in relation to itself, as Rose (1999) argues.

19 www.fastcompany.com/1840817/media6degrees-knows-what-you-want-buy-even-you-do (accessed February 20, 2016).

20 https://support.google.com/ads/answer/2662922?hl=en-GB (accessed November 15, 2015).

21 Ad blocking is a whole other sociomaterial assemblage which cannot be covered here. Refer to this source for a general discussion: www.computerworld.com/article/2487367/e-commerce/ad-blockers–a-solution-or-a-problem-.html (accessed February 20, 2016).

22 An honesty box is where a money box is provided and the participants are expected to place the right amount of money in the box for the goods that they take or consume. It is based on trust and the honesty of the participants.

23 From Google's help function "Autocomplete predictions are automatically generated by an algorithm without any human involvement, based on a number of objective factors, including how often past users have searched for a term" (https://support.google.com/websearch/answer/106230?hl=en, accessed February 20, 2016).

References

Arvidsson, A. 2005. *Brands: Meaning and Value in Media Culture*. London and New York: Routledge.

Baker, P. and Potts, A. 2013. "'Why Do White People Have Thin Lips?' Google and the Perpetuation of Stereotypes via Auto-complete Search Forms." *Critical Discourse Studies* 10 (2): 187–204. doi: 10.1080/17405904.2012.744320.

Barad, K. 2007. *Meeting the Universe Halfway: Quantum Physics and the Entanglement of Matter and Meaning*. Durham, NC and London: Duke University Press.

Barocas, S., Hood, S., and Ziewitz, M. 2013. "Governing Algorithms: A Provocation Piece." Paper for *Governing Algorithms Conference*, May 16–17, 2013, Rochester, NY. Retrieved from http://papers.ssrn.com/abstract=2245322 (accessed February 20, 2016).

Bateson, M., Nettle, D., and Roberts, G. 2006. "Cues of Being Watched Enhance Cooperation in a Real-World Setting." *Biology Letters* 2 (3): 412–414.

Butler, J. 1990. *Gender Trouble: Feminism and the Subversion of Identity*. New York: Routledge.

Çalışkan, K. and Callon, M. 2010. "Economization, Part 2: A Research Programme for the Study of Markets." *Economy and Society* 39 (1): 1–32.

Callon, M. 1986. "Some Elements of a Sociology of Translation: Domestication of the Scallops and the Fishermen of St Brieuc Bay." In *Power, Action, and Belief: A New Sociology of Knowledge?* Edited by J. Law, 196–233. London: Routledge & Kegan Paul.

Callon, M. and Muniesa, F. 2005. "Peripheral Vision: Economic Markets as Calculative Collective Devices." *Organization Studies* 26 (8): 1229–1250.

Castelluccia, C. 2012. "Behavioural Tracking on the Internet: A Technical Perspective." In *European Data Protection: In Good Health?* Edited by S. Gutwirth, R. Leenes, P. D. Hert, and Y. Poullet, 21–33. Dordrecht: Springer.

Chen, J., Liu, D., and Whinston, A. B. 2009. "Auctioning Keywords in Online Search." *Journal of Marketing* 73 (4): 125–141.

Curtis, S. 2014. "Would You Pay £140 a Year for an Ad-Free Web?" *Telegraph*, August 21. Retrieved from www.telegraph.co.uk/technology/news/11047801/Would-you-pay-140-a-year-for-an-ad-free-web.html (accessed February 20, 2016).

Dalessandro, B., Hook, R., Perlich, C., and Provost, F. 2012. "Evaluating and Optimizing Online Advertising: Forget the Click, but There Are Good Proxies." NYU Working Paper No. 2451/31637. Retrieved from http://papers.ssrn.com/abstract=2167606 (accessed February 20, 2016).

Evans, D. S. 2009. "The Online Advertising Industry: Economics, Evolution, and Privacy." *The Journal of Economic Perspectives* 23 (3): 37–60.

Foucault, M. 1977. *Discipline and Punish: The Birth of the Prison*. Translated by Alan Sheridan. New York: Vintage.

Foucault, M. 1978. *The History of Sexuality. Vol. 1*. Translated by R. Hurley. New York: Pantheon Books.

Foucault, M. 1980. *Power/Knowledge: Selected Interviews and Other Writings, 1972–1977*. New York: Pantheon Books.

Foucault, M. 1984. "Nietzsche, Genealogy, History." In *The Foucault Reader*. Edited by P. Rabinow, 87–90. Harmondsworth, UK: Penguin.

Foucault, M. 1991. "Governmentality." In *The Foucault Effect: Studies in Governmentality*. Edited by Graham Burchell, 87–104. Chicago, IL: University of Chicago Press.

Foucault, M. 2003. *The Birth of the Clinic* (Third Edition). Translated by Alan Sheridan. London and New York: Routledge.

Gillespie, T. 2014. "The Relevance of Algorithms." In *Media Technologies: Essays on Communication, Materiality, and Society*. Edited by T. Gillespie, P. J. Boczkowski, and K. A. Foot, 167–194. Cambridge, MA and London: MIT Press.

Goldfarb, A. and Tucker, C. 2014. "Standardization and the Effectiveness of Online Advertising." SSRN Scholarly Paper No. ID 1745645. Retrieved from http://papers. ssrn.com/abstract=1745645 (accessed February 20, 2016).

Hall, S. 1997. "The Work of Representation." In *Representation: Cultural Representations and Signifying Practices, Vol. 2*. Edited by S. Hall, 13–74. Thousand Oaks, CA: Sage.

Ingold, T. 2011. *Being Alive: Essays on Movement, Knowledge and Description*. London: Routledge.

Introna, L. D. 2006. "Maintaining the Reversibility of Foldings: Making the Ethics (Politics) of Information Technology Visible." *Ethics and Information Technology* 9 (1): 11–25.

Introna, L. D. and Ilharco, F. M. 2006. "On the Meaning of Screens: Towards a Phenomenological Account of Screenness." *Human Studies* 29 (1): 57–76.

Introna, L. D. and Nissenbaum, H. 2000. "Shaping the Web: Why the Politics of Search Engines Matters." *The Information Society* 16 (3): 169–185.

Introna, L. D. and Wood, D. 2004. "Picturing Algorithmic Surveillance: The Politics of Facial Recognition Systems." *Surveillance and Society* 2 (2/3): 177–198.

Jaworska, J. and Sydow, M. 2008. "Behavioral Targeting in On-Line Advertising: An Empirical Study." In *Web Information Systems Engineering: WISE 2008*. Edited by J. Bailey, D. Maier, K.-D. Schewe, B. Thalheim, and X. S. Wang, 62–76. Berlin and Heidelberg: Springer.

Kristol, D. M. 2001. "HTTP Cookies: Standards, Privacy, and Politics." *ACM Transactions on Internet Technology* 1 (2): 151–198.

Latour, B. 1988. *The Pasteurization of France*. Cambridge, MA: Harvard University Press.

Latour, B. 2005. *Reassembling the Social: An Introduction to Actor-Network-Theory*. Oxford: Oxford University Press.

Mitchell, R. 2014. "Ad Blockers: A Solution or a Problem?" *Computerworld*, January 15. Retrieved from www.computerworld.com/article/2487367/e-commerce/ad-blockers-a-solution-or-a-problem-.html (accessed September 29, 2014).

Nikiforakis, N., Kapravelos, A., Joosen, W., Kruegel, C., Piessens, F., and Vigna, G. 2014. "On the Workings and Current Practices of Web-Based Device Fingerprinting." *IEEE Security Privacy* 12 (3): 28–36.

Nissenbaum, H. and Brunton, F. 2013. "Political and Ethical Perspectives on Data Obfuscation." In *Privacy, Due Process and the Computational Turn: The Philosophy of Law Meets the Philosophy of Technology*. Edited by M. Hildebrandt and K. de Vries, 171–195. London: Routledge.

Orlikowski, W. J. and Scott, S. V. 2015. "Exploring Material-Discursive Practices." *Journal of Management Studies* 52 (2): 697–705.

Perlich, C., Dalessandro, B., Hook, R., Stitelman, O., Raeder, T., and Provost, F. 2012. "Bid Optimizing and Inventory Scoring in Targeted Online Advertising." In *Proceedings of the 18th ACM SIGKDD International Conference on Knowledge Discovery and Data Mining*, 804–812. New York: ACM.

Perlich, C., Dalessandro, B., Raeder, T., Stitelman, O., and Provost, F. 2014. "Machine Learning for Targeted Display Advertising: Transfer Learning in Action." *Machine Learning* 95 (1): 103–127.

Pickering, A. 1995. *The Mangle of Practice: Time, Agency, and Science*. Chicago, IL: University of Chicago Press.

Raeder, T., Stitelman, O., Dalessandro, B., Perlich, C., and Provost, F. 2012. "Design Principles of Massive, Robust Prediction Systems." In *Proceedings of the 18th ACM SIGKDD International Conference on Knowledge Discovery and Data Mining*, 1357–1365. New York: ACM.

Reimer, J. 2005. "A History of the GUI," May 5. Retrieved from http://arstechnica.com/features/2005/05/gui/ (accessed February 3, 2016).

Rose, N. 1999. *Governing the Soul: Shaping of the Private Self* (2nd Revised Edition). London and New York: Free Association Books.

Rose, N. and Miller, P. 2008. *Governing the Present: Administering Economic, Social and Personal Life*. Cambridge, UK: Polity Press.

Shah, R. C. and Kesan, J. P. 2009. "Recipes for Cookies: How Institutions Shape Communication Technologies." *New Media & Society* 11 (3): 315–336.

Skinner, D. 2013. "Foucault, Subjectivity and Ethics: Towards a Self-forming Subject." *Organization* 20 (6): 904–923.

Walgampaya, C. and Kantardzic, M. 2011. "Cracking the Smart ClickBot." In *2011 13th IEEE International Symposium on Web Systems Evolution (WSE)*, 125–134.

Whitehead, A. N. 1978. *Process and Reality: An Essay in Cosmology*. New York: Free Press.

Ziewitz, M. 2016. "Governing Algorithms Myth, Mess, and Methods." *Science, Technology & Human Values* 43 (1): 3–16.

Zuckerman, E. 2014. "The Internet's Original Sin." *The Atlantic*, August 14. Retrieved from www.theatlantic.com/technology/archive/2014/08/advertising-is-the-internets-original-sin/376041/ (accessed February 20, 2016).

3 #trendingistrending

When algorithms become culture

Tarleton Gillespie

To make sense of the increasingly complex information systems that now undergird so many social enterprises, some social scientists have turned their attention to the 'algorithms' that animate them. This "critical sociology of algorithms" (see Gillespie and Seaver 2015 for an evolving catalog of this work) has revived longstanding concerns about the automation and rationalization of human sociality, the potential for discrimination inside of bureaucratic and formulaic procedures, and the implications of sociotechnical systems for the practices that depend on them. Algorithms offer a powerful focal point for this line of inquiry: a hidden core inside these complex systems that appears to hold the secret, embedded values within. They are instructions, after all; the mechanic ghost in the machine? Tempting (Gillespie 2016; Ziewitz 2015). But, in our enthusiasm to install the algorithm as our new object of study, we (myself included) may have fallen into the most obvious of intellectual traps: the tendency to reify the very phenomenon we hope to explain. Much of this work positions 'the algorithm' as the thing to be explained, as the force acting on the world. This is hardly a new misstep; rather, it is one that has plagued the sociology of technology (Bimber 1994; Gillespie *et al.* 2014; Smith and Marx 1994; Sterne 2014; Wyatt 2008).

Invited to consider "algorithmic cultures," as we are in this volume, we might be tempted into the same trap: how has the introduction of algorithms changed the dynamics of culture? There are some interesting avenues to explore there, but they all run the same risk: of rehearsing a cause-and-effect story that treats 'the algorithm' as a single, bounded entity, presumes a stable and unsullied 'culture' that preceded this perturbation, and then looking for the effects of these algorithms on cultural practices and meanings—usually troubling ones.

But we will certainly come up short if we tell simple cautionary tales about the mechanisms of production and distribution and their effects, or reassuring fables about how they merely answer to the genuine wants of audiences. These are the intellectual missteps that plague the study of culture. Culture is the product of both of these corresponding, but not isomorphic, forces (Bourdieu 1993, 230). Cultural objects are designed in anticipation of the value people may find in them and the means by which they may circulate; once circulated, we encounter cultural objects amidst a corpus of others, and attend to their place amidst them (Mukerji and Schudson 1991). Moreover, culture is aware of this correspondence, self-aware and reflexive about its own construction. As we consume cultural objects, we sometimes wonder what it says about us that we consume them; and some cultural works are interested in culture itself, reading the popular as a clue to the society it is produced for and that finds meaning in it. Culture thinks about itself.

The mechanisms by which culture is produced and circulated are sometimes drawn up into those debates, and the signals of valuation (Helgesson and Muniesa 2013) they generate—of what is significant, or popular, or newsworthy, or interesting—themselves become points of cultural interest, telling us something about the 'us' to which it is served. We not only debate the news item that made the front page, we sometimes debate the fact that it made the front page, the claim of importance made by the newspaper in putting it there, the logic by which newspapers choose and prioritize news stories, the institutional forces that drive modern news production. Evidence that we want a particular cultural artifact, or claims that we should, provoke us to ask why: why is this particular cultural object popular, how did it become so, are the artists and industries that helped generate it feeding us well, should culture be popular or should it be enlightening, are other kinds of culture being displaced in the process?

Today, these questions have algorithms in their sights, particularly those algorithms that help select and deliver the cultural works we encounter. Algorithms, particularly those involved in the movement of culture, are mechanisms of both distribution *and* valuation, part of the process by which knowledge institutions circulate *and* evaluate information, the process by which new media industries provide *and* sort culture. In particular, assertions of cultural value, always based on prediction, recipes, and measurements of what makes something culturally valuable, are incorporating algorithmic techniques for doing so. Algorithms, then, are not invisible. While they may be black boxes in terms of their code, at the same time they are often objects of public scrutiny and debate.

Not only should we ask, then, what it means for modern culture industries to adopt algorithmic techniques for producing, organizing, and selecting culture, and for knowing, tracking, and parsing audiences in order to make those choices (Beer and Burrows 2013); these are deeply important questions. But we should also examine the way these algorithmic techniques themselves become cultural objects, get taken up in our thinking about culture and the public to which it is addressed, and get contested both for what they do and what they reveal (Striphas 2015). We should ask not just how algorithms shape culture, but how they become culture.

Trending algorithms and how they work

Given their scale, all social media platforms must provide mechanisms to 'surface' new and relevant content, both to offer the user points of entry into and avenues through the archive, and thereby to keep users on the site longer, exposing them to more ads and collecting more valuable data about them. Most navigation mechanisms are either *search* or *recommendation*: search, where the user's query is matched with the available content; and recommendation, where the user is matched with other users and offered content they preferred. (Many navigation mechanisms are now a combination of the two; I'm separating them here only for analytical purposes.)

If not as pervasive or structurally central as search and recommendation, *trending* has emerged as an increasingly common feature of such interfaces and seems to be growing in cultural importance. It represents a fundamentally different logic for how to navigate social media algorithmically: besides identifying and highlighting what might be relevant to 'you' specifically, trending algorithms identify what is popular with 'us' broadly. The simplest version of trending is as old as social media: using some simple measure of recent activity across a site to populate the front page with popular content. More sophisticated techniques, what I will broadly call *trending algorithms*, use a combination of metrics to identify particular content or topics generating the most activity, at a particular moment, and among a particular group of users.

One of the earliest and most widely known of these is Twitter Trends, introduced in September 2008 (though Google introduced its Trends Lab back in 2006, before Twitter launched). Twitter Trends is a brief list of the hashtags and other terms that are appearing more than usual at that moment, specific to the user's city (within the U.S.) or country (see Figure 3.1). The terms are ranked and, if clicked, conduct a search on that term, presenting the user with the most recent uses of it.

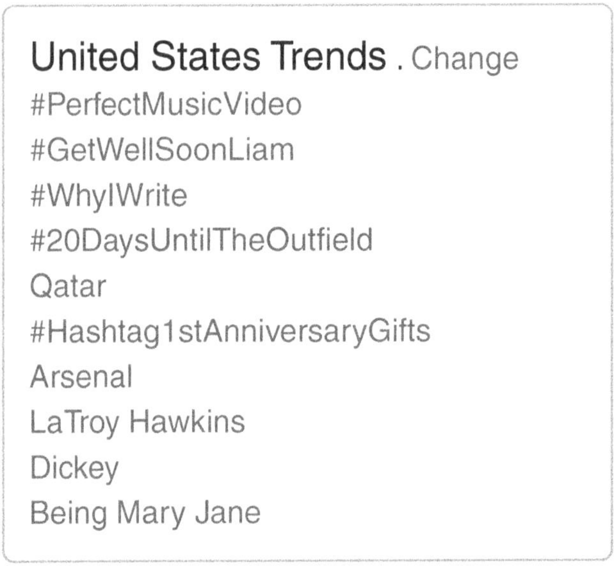

United States Trends . Change

#PerfectMusicVideo

#GetWellSoonLiam

#WhyIWrite

#20DaysUntilTheOutfield

Qatar

#Hashtag1stAnniversaryGifts

Arsenal

LaTroy Hawkins

Dickey

Being Mary Jane

Figure 3.1 Twitter Trends (© Twitter, 2015).

By indicating that "Arsenal" is trending, the algorithm indicates that lots of people seem to be using the phrase in their tweets, more so than usual, enough to stand out above all other topics. It implies that a group of people (a public if you will, though a momentary one) has gathered around a common interest. Some trending topics are referential to phenomena beyond Twitter, like a beloved sports team or politically relevant event, while others may index discussions that emerge on Twitter exclusively, like "#PerfectMusicVideo." What puts them on the list is an algorithmic calculation, the details of which are largely opaque to the user.

Twitter Trends may seem like a minor feature. The list occupies a relatively small box in the lower left-hand corner of a Twitter user's homepage; for those accessing Twitter on their mobile phone, Trends were only recently added, displayed only when the user first initiates a search but before beginning to type. For users who access Twitter through a third-party app, Trends may be harder to locate or even be unavailable. It is also easy to discount, often full of gimmicky hashtags, pop culture fads, and seemingly meaningless terms. Nevertheless, it is a means by which users find their way to new topics, one of what Bucher (2012) calls the "technicities of attention" that social media interfaces provide. To the extent that it surfaces content, it elevates its visibility and directs users toward it, at least potentially.

Trending algorithms calculate the current activity around every post, image, or hashtag, by combining some measure of each item's popularity, novelty, and timeliness.[1] Within these measures are a number of assumptions. In particular, trending algorithms tend to be concerned with a very broad *who*, and a very narrow *when* (and a little bit about *what*).

Who: Trending algorithms start with a measure of popularity, for instance how many users are favoriting a particular image or using a particular hashtag. But this entails deciding first who counts. Is it all users on the platform, or a subset? They are often bounded regionally (only counting U.S. users, for example); this may be presented as a feature of the results (as with Twitter), or it may remain obscure within the calculation, leaving it to the user to imagine who the 'us' is. Platforms sometimes also factor in information about the users whose activity they are assessing, such as weighing the activity of popular users more heavily, or factoring in whether the popularity of an item is bounded within or spans across clusters of users already linked as friends or followers.

When: Trending algorithms emphasize novelty and timeliness, both in terms of identifying unprecedented surges of activity, and in aspiring to match those to real events happening now. The parameters of what 'now' means are often oblique: are these items popular over the last minute? hour? day? In addition, to identify a surge of activity requires a baseline measure of what's typical for this item. This usually means selecting a point in the past for comparison: how much more discussion of this topic is there now, compared to the same hour one week ago? This can require tricky mathematical adjustments, to compensate for topics that have very little activity (if a topic had one mention last week and two mentions this week, is that an enormous jump in activity or a meager one?) or for topics that have no precedent with which to compare (the first discussion of a new movie title, or a viral hashtag in its first appearance).

(*What*: Trending algorithms are almost entirely agnostic about what content they are identifying. They must exclude words that are too common to trend: something like "today" probably shouldn't be there, although if its use surged over its typical usage, perhaps something different is happening? They must also discern when the same word has different meanings: is it "today" as in, say, what is current, or as in the NBC morning news show? And they must recognize when different terms should be counted together: perhaps "Today Show" and "Today" and #todayshow and #today should be considered a single instance. All of this depends on a great deal of machine learning and rough guesswork. And platforms adjust their trending algorithms to better approximate the kind of results they want.[2] It is also worth noting that Twitter Trends tries to exclude profanity, terms identifiable as hate speech, and other obscenities—regardless of their popularity.[3] Many other social media do the same.)

This means, of course, there are different ways to make these calculations. Reddit, for example, offers several trending algorithms for identifying what's popular, including "new," "rising," "controversial," and "top." Factoring the who, the when, and the what in different ways, or including other factors, generates different results.

Trending algorithms beyond Twitter

Twitter Trends has enjoyed the most visibility. But we should not be misled by the relative prominence of Twitter's version, or the current zeitgeist of the term 'trending' itself. I want to broaden the category of trending algorithms to include the myriad ways in which platforms offer quick, calculated glimpses of what 'we' are looking at and talking about.

Most social media platforms now offer some measure of recent and popular activity. Similar Trends lists now appear on Facebook, Instagram, YouTube, Tumblr, Pinterest, and Vine. Reddit's front page is organized as a list of recent posts ranked according to how users have upvoted and downvoted them, with decay over time factored in—a similar aggregation of the popular and the current. Google and Bing search engines offer Google Trends and Bing's "Popular Now" that digest the most popular search queries of the moment. Apple's App Store lists "trending searches" before the search bar is filled in; Etsy will email you the latest trends among their offerings. Many news and entertainment sites offer similar mechanisms: The *New York Times*, just as one example, highlights articles most frequently viewed or emailed by readers and those most shared on Facebook, based on a measure of the previous 24 hours of activity on the site (see Figure 3.2). Hulu includes "what's trending now" as one of its navigation menus.

Some social media platforms issue "trend reports," not in real time but at particular moments. These include "this year in trends," such as those produced by Google and Bing, that use the most popular search queries to craft a visual reminiscence of the cultural highlights of the past year. Other sites publish glimpses

The New York Times

Trending

Good evening. Catch up on the stories resonating with our readers this minute.

At Stanford, an Affair Reveals
Accusations of Discrimination

For Canada, Justin Trudeau
Election Offers Change of Tone

The Strange Case of Anna Stubblefield

She told the family of a severely disabled man that she could help him to
communicate with the outside world. The relationship that followed
would lead to a criminal trial.

The Lonely Death of George
Bell

Brazil Pension Crisis Mounts
as More Retire Earlier, Then
Pass Benefits On

What Do We Really Know
About Osama bin Laden's
Death?

Your Tuesday Evening
Briefing: Justin Trudeau, Oscar
Pistorius, Syria

Consumer Reports Stops
Recommending the Tesla

American Cancer Society, in a
Shift, Recommends Fewer
Mammograms

Figure 3.2 "Trending" (© *New York Times*, 2015).

of their data, as blog posts or infographics, revealing something about popular activity and taste on their site (while also showing off their capacity for data analytics). OK Cupid, Foursquare, Spotify music streaming, and adult video platform Pornhub have been notable in serving up these glimpses of what is most popular on their site, producing analyses and infographics that then circulate on sites like Buzzfeed: what men or women rate highly in their partners, the most popular porn search terms by state (see Figure 3.3), surges in site activity after a major national event, or what songs are most distinctly popular in a particular city this month. OK Cupid's founder even turned his site's data analytics into the book *Dataclysm*, with the provocative subtitle *Who We Are (When We Think No One's Looking)* (Rudder 2014). While these are more retroactive synopses than real-time snapshots, like other trending algorithms they aggregate some subset of activity on their platform over some specific time parameter, and constitute that data into representations of popular activity and preference.

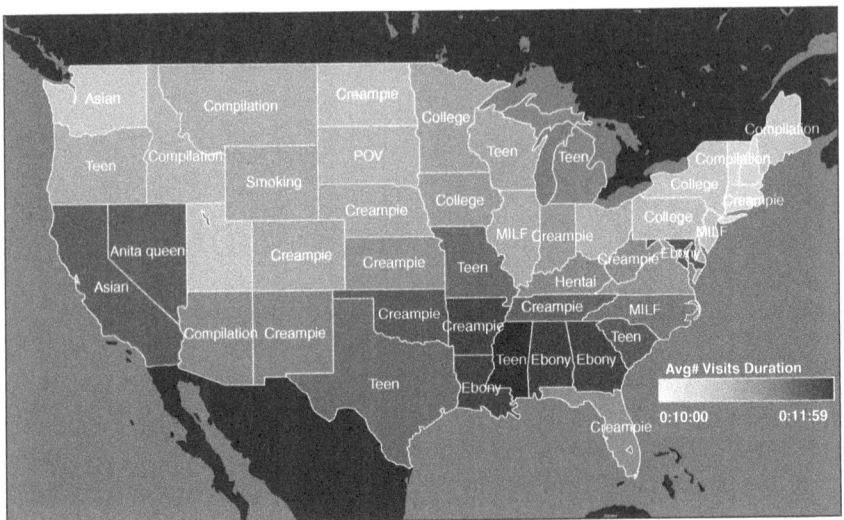

Figure 3.3 "Pornhub's US Top 3 Search Terms by State" (© Pornhub, 2013).

E-commerce sites such as Amazon list the sales ranks of their products. On first blush, these may not seem to belong in the same category as trends, as they claim to measure a much more straightforward data point: number of purchases of a given product among all products. But it is a very similar mechanism: a measure of popular activity, bounded in oblique ways by timeframe, category, and other parameters determined by the platform, and fed back not just as information but as an invitation to value that product because of its popularity.

Sales ranking also does not include everything: Amazon's is carefully moderated for inappropriate content, just as most trending algorithms are. This was made apparent by the "Amazonfail" incident, when thousands of gay and lesbian fiction titles temporarily lost their sales rank because they had been incorrectly classified as "adult."[4] This is a small but important reminder that, like other trending algorithms, sales rank is a public-facing representation of popularity, not just a pure tabulation of transactions.

Let's also include navigation tools that may feel somewhat more incidental, but nevertheless are legible as glimpses of popular activity. Consider Google's autocomplete function, where the site anticipates the search query you're typing based on the first few letters or words, by comparing it to the corpus of past search queries. While the primary purpose of autocomplete is merely to relieve the user of typing the remainder of their query, the suggestions it makes are a kind of measure of popular activity and taste (at least as represented through searching on Google).

It appears we are awash in these algorithmic glimpses of the popular, tiny barometers of public interest and preference produced for us on the basis of platform-specific activity, inviting us to both attend to and join these trends.

Moreover, the word 'trending' has escaped Twitter and its competitors, showing up across cultural, advertising, and journalistic discourse. It is an increasingly common trope in ad copy, fashion spreads, news reports, even academic publishing (see Figure 3.4).

This is not to suggest that advertisers and news programs have never before wanted to get our attention by telling us what's popular. But the fact that the term 'trending' is enjoying a zeitgeist moment is indicative of the way that this particular formation of popularity has captured our attention and imagination.

The effects of trending algorithms?

Search was the first point of concern for sociologists interested in algorithmic media. Whether or not they used the term 'algorithm,' investigations into the possible biases of search (Granka 2010; Halavais 2008; Introna and Nissenbaum 2000) and the personalization of news (Pariser 2012) were concerns about algorithms and their impact on culture. What animated that work was the disappearance of

Figure 3.4 "What's Trending?" (Routledge promotional email, © Taylor & Francis, 2014).

common experience and the fracture of publics, and the growing privacy abuses and information exploitation as platforms sought more ways to know the preferences of each individual user (Stalder and Mayer 2009; Zimmer 2008).

Unlike search, trending algorithms promise a glimpse into what may be popular or of interest to others, a barometer of 'what's going on.' They offer the kind of serendipity that the personalized news environment threatened to do away with. They call together publics rather than fracturing them (while privileging some publics over others).

On the other hand, they are not so much glimpses as they are hieroglyphs. 'Trending' is an oblique category these measures rarely unpack. Trends are not independent phenomena: unlike something like the number of subscribers or the number of likes, they do not even claim to represent verifiable facts. 'Trends' could mean a hundred things, and may mean almost nothing. Trending algorithms don't even have to be right, in the strictest sense; they merely have to be not wrong. But they do aspire to say something about public attention, beyond the user-selected community of friends or followers; they say something—perhaps implicitly, perhaps incorrectly—about cultural relevance, or at least we are invited to read them that way. They crystallize popular activity into something legible, and then feed it back to us, often at the very moment of further activity.

Scholars interested in social media platforms and particularly the algorithms that animate them have begun to think about the importance of metrics like Twitter Trends, and more broadly about the "metrification" of social activity online (Beer 2015; Beer and Burrows 2013; Gerlitz and Lury 2014; Grosser 2014; Hallinan and Striphas 2014; Hanrahan 2013; Lotan 2015; Marwick 2015). First, there are important questions to ask about how these measures are made and how they shape digital culture. Similar questions have been raised about the measurement of public opinion (Beniger 1992; Bourdieu 1972; Herbst 2001). How are claims of what is 'trending' reached, who do they measure, and how might these claims be biased? The computational and statistical procedures used to assess popular taste may be biased in particular ways (Baym 2013). Trends may measure some kinds of audience activity better than others, not only overlooking other popular activity but further rendering it seemingly irrelevant. And, as only a few institutions can generate these metrics at scale, and many of them are the producers and distributors (and platforms) themselves, there is a risk of self-serving biases, to form the kinds of collectivities they hope to produce and cater to with their platform.

Second, what are the effects of these metrics when delivered back to audiences? There is evidence that metrics not only describe popularity, they also amplify it, a Matthew Effect with real economic consequences for the winners and losers. Some social media platforms are structured to reward popularity with visibility, listing the highest-ranking search results or the content voted up by a user community nearest to the top of the page. If visibility matters for further exposure, then the metrics of popularity that determine visibility matter as well (Sorenson 2007; Stoddard 2015). Further, some consumers use metrics as a proxy for quality: buying from the bestseller list or downloading the most downloaded song is a better strategy than random for getting something good. This means early winners can see that popularity compounded (Salganik and Watts 2008).

The dynamics of these feedback loops are likely to be more pronounced and intertwined for trending algorithms. Because the calculation is in near real-time (Weltevrede *et al.* 2014), and is fed back to users at exactly the point at which they can interact with that highlighted content, the amplification of the popular is likely heightened. As David Beer (2015, 2) has noted, we are seeing "an emergent 'politics of circulation' in which data have come to recursively fold-back into and reconfigure culture." In some cases, these are central to the platform's interface. For instance, click on an artist in Spotify, and the first and most prominent offer is that artist's top five songs, according to Spotify's measure of play count, though adjusted for how recent the music is—in other words, a trending algorithm. These five songs are not only more likely to be played, they are presented as a way to encounter and evaluate that artist. Furthermore, Trends are self-affirming: click a trending topic on Twitter and you immediately enter a discussion already underway, visceral proof that the topic is popular (regardless of what other topics may in fact exceed it, or what kind of populations are or are not helping that topic trend).

Moreover, because trending algorithms attend to such a broad *who* and such a narrow *when*, their shape could affect the temporal qualities of cultural discourse. It is not new to suggest that popular culture, especially in the West, has become ever more concerned with speed. News cycles, the rapidity with which hit movies or popular songs come and go, and the virality of digital culture, all suggest that contemporary culture is more interested in timeliness and novelty. The effort to get a topic to trend means playing the game, of breadth and speed, getting a discussion to surge in exactly the right way for a trending algorithm to recognize it. We may see something similar to the emergence of the "sound bite" (Hallin 1992), a similar shaping of cultural practices oriented towards capturing the attention of news institutions obsessed with brevity. Powers (2015) makes a similar point in her discussion of "firsts"—when online commenters try to post first in a thread, regardless of whether they have anything to contribute. This particular "metaculture" (Urban 2001, quoted in Powers 2015), or the cultural shape of culture, is a complex combination of being first in time, first on a list, and first as best—a combination that unites other structures like 'top 10' or 'breaking news' or 'soundbite.' It is a combination that 'trending' shares as well.

As Beer and Burrows (2013) observe,

> This is a much accelerated and in some ways entirely new set of organizations and relations in popular culture about which we so far have little understanding. Nor, we could add, do we have a clear sense of the socio-technological infrastructures and archives that organize and underpin it, the way the data is played with or algorithmically sorted, and how this shapes culture.
>
> (Beer and Burrows 2013, 65)

Knowing the popular, from tastemakers to audience metrics to infomediaries

But trending algorithms, while they may be new in the technical sense, are not new in the cultural sense. They build on a century-long exercise by media industries to identify (and often quantify) what's popular, and they innovate the ways in which these measures themselves feed back into cultural circulation. We have long encountered culture through both subjective assertions and manufactured metrics about what's popular, what's about to be, and what should be. This means that trending algorithms and their role in social media platforms must be understood in light of their twentieth-century analogues.

Here I am linking the study of algorithms to the broader interrogation of the material, institutional, and economic structures of media, and the implications those structures have for the circulation of culture (Williams 1958). This includes attention to the political economy of cultural production and distribution (Garnham 1979; Mansell 2004), and specifically the commercial industries that dominate those undertakings (Havens *et al.* 2009; Jenkins 2006; McChesney 2015), and how social media platforms increasingly play that role (Burgess and Green 2013; Gillespie 2010; Napoli 2015; Sandvig 2015; Vaidhyanathan 2012; van Dijck 2013); the dynamics of cultural production that shape content (Braun 2015a, 2015b; Holt and Perren 2009; Peterson and Anand 2004), and the work of "cultural intermediaries" that facilitate and legitimate the making of symbolic goods and meaning (Bourdieu 1993; Downey 2014; Hesmondhalgh 2006; Neff 2012). The link is made most explicitly by Morris (2015) in his discussion of "infomediaries," where he considers the work of algorithmic curators (his example is music recommendation systems) as analogous in important ways to the (human) cultural intermediaries that concerned Bourdieu.

Like information media, the nineteenth- and twentieth-century media industries, from book and magazine publishing to broadcasting to the distribution of music and film, were dependent on the economics of 'public goods' where initial costs are high and distribution costs are relatively low, and on the anticipation of the fickle tastes of audiences. As they grew in scale and ambition, they sought ways to make popular taste legible and to deliver those preferences back to audiences.

Producers and distributors eager to anticipate and shape popular tastes turned first to subjective and impressionistic tastemakers: disc jockeys, book reviewers and film critics, and cultural commentators. These evaluators of the popular depended on a combination of personal or subjective acumen, expertise, and some purported or demonstrable capacity for taking the public's pulse. The fact that Twitter and other social media platforms called their mechanisms "trends" harkens back to this tradition of cultural tastemaking: magazines devoted to identifying trends in fashion, DJs with an ear for emerging music genres, industry executives with an intuitive sense for 'the next big thing.' Today, bloggers, podcasters, makers of playlists—and maybe all of us (Maguire and Matthews 2012)—are carrying the role of cultural intermediary into the twenty-first century.

For the media industries, reading the public and anticipating its wants in this haphazard way appeared increasingly insufficient and risky. However, they did have another way to evaluate what was popular, at least around their own products: "Simultaneously with the development of mass communication by the turn of the century came what might be called mass feedback technologies" (Beniger 1989, 376). These companies sought increasingly sophisticated versions of sales data, including circulation numbers for newspapers and magazines, box office receipts, and audience ratings for radio and television (Napoli 2003). Some of this was part of the industrialization of the distribution process itself, entries on a ledger for how newspapers or LPs moved from warehouse to shop counter. Advertisers in particular wanted more information about the audiences they were buying, more objectively acquired, and at greater granularity (Ang 1991).

But this was no small undertaking, requiring decades of innovation for how to track sales and circulation data on a national and even global scale, and how to make sense of that data according to demographics, region, and genre. The first measures were clumsy—as both Ang (1991) and Napoli (2003) note, early radio stations would weigh fan mail as a rough assessment of popularity. These early and blunt feedback mechanisms were increasingly replaced by more rationalized, analytical approaches to understanding public behavior and taste, the emergence of new professionals and disciplines (like psychology) for audience measurement, and eventually the rise of third-party services like Nielsen for tracking audience data and selling it back to the industries who wanted it (Napoli 2003). These claims, while more grounded in data of actual consumption, cannot entirely shed the more impressionistic quality of tastemakers, as they hope to identify and generate surges of popularity as or before they crest.

This turn to audience metrics represented a transformation of the media industries and the cultural work they helped to circulate, as both audiences and even products came to be understood in terms of these metrics of popularity (Napoli 2003). But the rise of audience data was concomitant with a broader embrace of and fascination with surveys and other measurable social data in the world at large (Igo 2008). In large-scale projects like the Lynd's Middletown study, political polling by Gallup and others, and the studies of human sexuality by Kinsey, large-scale and quantifiable social science research techniques were used to satisfy an emerging interest in both the typical and the aggregate. Alongside these projects, market research and media industry audience research took up these same tools to ask similar questions. The current public interest, not only in 'trends' but in infographics, heat maps, and forecasts of online activity and cultural preference, is part of a century-long fascination with social data and what it promises to reveal about the public at large.

With the shift to digital production and distribution, a radical new scale of data about audience activity and preference can be collected, whether by content producers, distribution platforms, or search engines. The digestion and exploitation of this data is a fundamental process for information intermediaries. Trending algorithms have become a structural element of social media platforms in part because they are a relatively easy and incidental bit of data for platforms to serve back to users. We might think of trends as a user-facing tip of an immense back-end iceberg, the enormous amount of user analytics run by platforms for their own benefit and for the benefit of advertisers and partners, the results of which users rarely see.

Morris suggests that we think more about "infomediaries": "an emerging layer of organizations ... that monitor, mine and mediate the use of digital cultural products (e.g., e-Books, music files, video streams, etc.) as well as audience responses to those products via social and new media technologies" (Morris 2015, 447). These infomediaries have taken up the role of both tastemaking and "audience manufacture" (Bermejo 2009). Further, we are beginning to see the automatic production of information, generated on demand in response to the measure of public interests (Anderson 2011).

Situating trending algorithms as part of a historical lineage of efforts to "know the popular" highlights some interesting features about trending algorithms and how they mediate our engagement with culture. They are part of a much longer debate about how culture is produced and measured; how those metrics are always both mathematical and subjective, always shaped by how they are measured by and bent through the prism of commerce; and how those measures are made meaningful by the industries and platforms that generate them.

Metrics become cultural objects themselves

Still, if we think about trending algorithms only in terms of their possible impact, I would argue, we miss an important additional dimension: the way they quickly become cultural objects themselves. They matter not only because they represent culture in particular ways, and are acted upon with particular consequences; they matter also because they come to be culturally meaningful: points of interest, 'data' to be debated or tracked, legible signifiers of shifting public taste or a culture gone mad, depending on the observer. When CNN discusses "what's trending" on Twitter it is using Trends as an index of the popular, and treating that index as culturally relevant. Measures of what's popular tell stories about the public, and are made to tell stories by those who generate and attend to them.

Once again, audience metrics are a useful point of comparison. In the second half of the twentieth century, audience metrics were not only consumed by industry professionals, but by the broader public as well. They were incorporated into advertising—"Number 1 at the box office!"—and circulated more broadly as part of an entertainment press reporting on popular entertainment. Newspaper sections devoted to books or movies began to report the bestseller lists and the weekend box office returns, covering the week's winners much like they cover elections. Trade magazines that cover specific industries, like *Billboard* and *Variety*, have expanded to increasingly address non-industry audiences; popular magazines like *TV Guide*, *Entertainment Weekly*, and *Rolling Stone* report ratings and sales data alongside their articles and reviews. Increasingly, part of being a media fan is knowing how much money a movie made in its opening weekend, which show won its time slot, or what album had the biggest debut.

Perhaps the most striking example is the long-running radio program *American Top 40*, hosted by Casey Kasem. Building on the emergence of "Top 40" radio stations devoted to playing only the most popular hits, the program's conceit was to play the 40 most popular songs in the U.S. that week, based on data from *Billboard*'s "Hot 100" singles chart. The show was quickly embraced—beginning on July 4, 1970 on seven stations in the U.S., at its most popular it was syndicated on more than 1,000 stations in over 50 countries. For the next few decades it had an outsized influence on American music culture. Before MTV, digital downloads, or streaming music services, it was one of the few places to hear the most popular music in the country. And it offered listeners the pleasure of tracking which songs were rising and falling in popularity, and which would be crowned number one for that week (Weisbard 2014).

This was not the first time that music sales were turned back to the audience as a contribution to culture. Local radio stations had begun to broadcast countdowns, though their measure was limited to the station's regional audience. *American Top 40*'s most direct predecessor, *Your Hit Parade*, ran from 1935 to 1953 as a radio program and through the 1950s as a television show, broadcast studio musicians and vocalists performing the most popular songs nationwide. *Your Hit Parade* was more circumspect about exactly how this popularity was determined—based on an 'authentic tabulation' of surveys of U.S. listeners, jukebox providers, and sheet music sellers, conducted by American Tobacco, the show's sponsor.

Billboard magazine itself had been in print since 1894, originally tracking outdoor advertising and amusements before expanding to film, vaudeville, and live entertainment. Charts for sheet music sales appeared in the 1910s, 'hit parades' that tracked the most popular songs on U.S. jukeboxes were added in the 1930s, followed by charts for broadcast music in the 1950s. Though *Billboard* was available to individual subscribers, it was intended as a trade magazine for advertising and music professionals; it was *American Top 40* that turned its metrics out for popular consumption.

Besides broadcasting the results of *Billboard*'s measurements, *American Top 40* became a cultural icon in and of itself. It offered a ritual for music fans in the 1970s, 1980s, and 1990s, a shared media text. Kasem became a well-known celebrity, and many of the details of the show became widely recognized cultural touchstones: the sonic flourishes used to bring the show back from commercials, the 'long distance dedications,' and Kasem's sign-off phrase. *American Top 40* was culturally meaningful, not only for the artists (whose popularity and income was deeply affected by 'charting' or reaching number 1) but for fans who listened, some of whom studiously kept track of the shifting fortunes of favorite artists, argued why this or that artist should or should not have made the list, aspired to get their own comments on the air. Providing an accurate report on the tastes of the American public was only part of its popularity. It was American music's version of Trends, with *Billboard* as its trending algorithm.

Metrics can become an object of cultural concern

As long as *American Top 40* was on the air, and long after, people debated the show and the vision of music and the American public that it offered. Even Kasem's death in 2014 revived debates about the populism and artifactuality of his show and its effect on U.S. music culture.[5] Was it a center point, uniting audiences around the most shared and beloved music of the moment? Or was it the product of an already narrow radio formatting too focused on hits? Was it meritocratic, introducing new performers and challenging musical forms despite the conservatism of radio programming? Or did it further marginalize genres like hip hop, metal, and country, categories often associated with working-class and black audiences? Did it make commerce the predominant metric for measuring the value of music? Or did it listen to fans, better than the market could? Did it represent a 'shared culture,' around the likes of 1980s superstars like Michael Jackson and Madonna, or was this 'shared culture' merely an artificial construct of the *Billboard* charts and the show itself, that crumbled in the face of the fragmentation of music in the 1990s?

Debates about the nature and value of the 'popular' in culture both predate and extend past *American Top 40*. What does 'popular' mean in the vocabulary of each of these metrics, and what should it mean? Does the amplification of the popular do harm or good to the culture? Such concerns implicate *American Top 40*, the *Billboard* charts, the historic formats of American radio, and the structure of the music industry itself. What does it mean that commercial mechanisms measure and make claims about the popular? When information intermediaries offer us the popular, is that a reflection of our wants or the manufacture of them?

We can hear similar debates today, about social media platforms and what they're for, about how social media amplify the popular back to us and with what effect. As long as social media have existed, we have debated whether they convey information of sufficient importance. The well-worn critique of Twitter, "I don't care what my friend had for breakfast," echoes early critics of radio. Similar laments about the effect of social media on journalism suggest that the political and civic-minded will be drowned out by narcissism and frivolity. Criticizing Facebook's algorithm, Tufekci (2015) wondered what it meant that, while her Twitter feed was full of news and comment about the protests in Ferguson, her Facebook feed was, dominated instead by videos of the "ice bucket challenge." What if what we want is precisely our downfall?

Sometimes trending algorithms play a part in those contestations, and are sometimes even redesigned in the wake of such debates. Twitter Trends offers an opportunity to debate what appears there—or, more importantly, what does not. Some say it celebrates pop culture trash and silly hashtag games; others have called it out for overlooking important topics. I have written elsewhere (Gillespie 2012) about the concerns raised by political activists, both around the Occupy Wall Street protests and the classified documents published by WikiLeaks, when a seemingly popular term fails to Trend. Charges of "censorship" overshadow more complex questions about the workings of trending algorithms, how they measure popularity, and what assumptions users make about what does and does not appear there. But they resonate because Trends is a powerful and consequential measure of the popular, and is often taken to be so in the wider culture. Occupy critics may have been wrong about why their hashtag didn't Trend, but if CNN and the wider culture assumes that trends = importance, they were not wrong for worrying.

Metrics can provide a venue to think about ourselves as a public

Measures of the popular claim to represent the public and its tastes, though it might be more accurate to say that they momentarily bring a 'public' into being around this claim of shared preferences. As Raymond Williams noted, "there are in fact no masses, but only ways of seeing people as masses" (Williams 1958, 11; cited in Baym). But whether we think of these metrics as reflections of a public or as constituting one, they certainly are often taken as revealing something about that public, by both industry insiders and listeners. A public is brought into focus, made legible; a listener of *American Top 40* feels like they know something about their fellow listeners, and about the culture of which they themselves are a part.

Social media algorithms generate "calculated publics" (Gillespie 2014): they imply a body of people who have been measured or assessed, as an explanation for why particular information has been presented as relevant to them. This is true for search and recommendation, and it is true for trending as well: when search results are returned to our query, there is some undefined population of users who have found these sites relevant, and have left their assessment in traces like incoming links and past clicks. When a movie is recommended based on "people like you," users enjoy a passing glance of a public to which they apparently belong, though one they can only know through the algorithmic results delivered on that basis. Trending algorithms make the claim of this calculated public more explicit: this is what 'we' are reading, this is what my city or country is tweeting about, this is what America is listening to today.

Who and when this public is, exactly, is less clear. While *American Top 40* explicitly stated that it based its rankings on the *Billboard* charts, any specific knowledge about how *Billboard* produced those charts was not conveyed, nor was it important to the experience of the program. You need not know how *Billboard* measures music sales to enjoy the countdown.

But while who was being measured was left unclear, the program told us who it was about and for, again and again, as part of its performance. Even from the program's name and logos, red white and blue and festooned with stars, it was clear that this public was an American one.

The program also performed 'America' as its spatial imaginary throughout the show, from the recurring tagline "The hits from coast to coast!" to Kasem welcoming new affiliate radio stations by their city and state (and regularly highlighting that show proudly appeared on American Armed Forces Radio), to his interstitial flourishes like "from the rocky coasts of Maine to the sandy shores of Hawaii." The program was not just listing popular songs of the moment, it was performing America itself. Any mismatch between 'America' and who was actually tabulated in *Billboard*'s charts was completely elided.

Twitter Trends indicates what region is being measured; I might choose "Boston Trends" or "United States Trends" or any region that Twitter offers, whether I live there or not. The specifics of how Twitter bounds these places, or these sets of geo-located users, are left unspecified. But for many trending algorithms, American-ness is assumed, or offered as the default. *American Top 40*'s emphasis on America may be more like the trends infographics that gather search data from Pornhub or check-ins from Foursquare, which always seem to cast it back on the familiar outline of the 50 states—an intuitive and conventional way to make sense of shared preferences, whether state lines have anything to do with the commonalities of cultural meaning or the flow of online discussion being represented.

But the fact that the 'us' being measured is left vague also means it can be contested. Trends results can become a terrain for discussion about who gets to be in the public square, how their visibility matters, and what happens when competing publics collide in a single space. UN Women used this to great effect in the 2014 campaign "The Autocomplete Truth" (produced with ad agency Memac Ogilvy & Mather Dubai), intended to raise awareness about violence against women, by showing the reprehensible Google autocomplete results to sentence fragments like "women shouldn't…" (see Figure 3.5). In tiny print

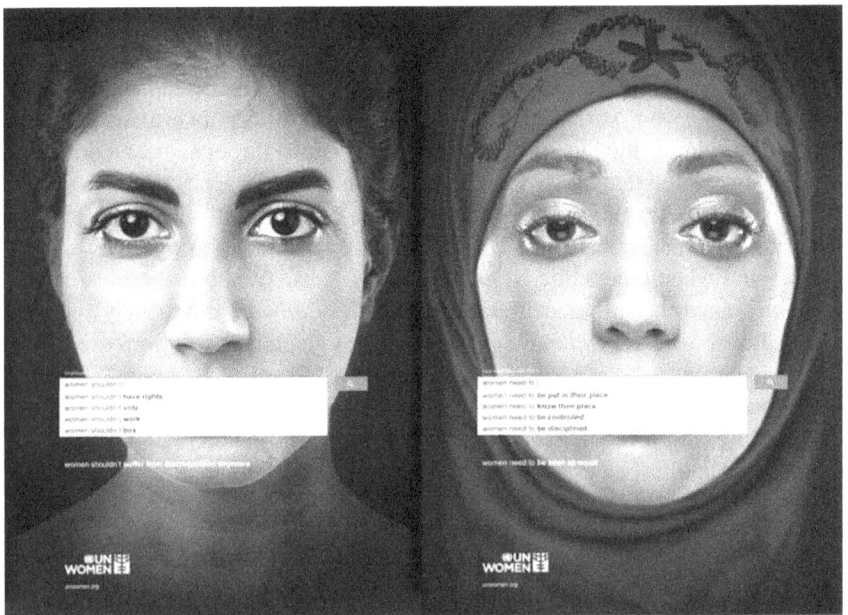

Figure 3.5 "The Autocomplete Truth" (© UN Women, 2013).

each poster asserted "Actual Google search on 08/03/2013." The message is a sobering one; its power depends on the presumption that the autocomplete algorithm reveals what 'people' really think, or at least really search for—"who we are, when we think no one's looking."

Particularly tricky discussions have erupted around the visibility of race, and a subpopulation of Twitter users commonly referred to as "Black Twitter." The topics important to this community will only sometimes reach the thresholds sufficient to be recognized by Twitter's algorithm; when they do, they have elicited xenophobic reactions.[6] The "what are these topics doing here" outcry rests on presumptions of who the "here" belongs to, and what happens when the measures suggest otherwise. There, reactions echo the panics around *American Top 40* when hip hop began to chart alongside white artists. The very offer of a common space in which popularity amidst an entire public will be represented, even when the contours of that 'entirety' are opaque and contested, can make terrain for debates about who is in that public, competing subcommunities with a single public, and who and what deserves representation there.

Conclusion: when algorithms become culture

Trending algorithms measure, and they also announce. This makes them data-based and calculating, and in doing so, they offer up a rich hieroglyph about some 'us,' some public, that can itself be discussed, marveled over, or rejected, just like finding out that some crappy pop group just took the #1 spot from your beloved indie band. They can be cultural objects of meaning, not just for those producing information and looking to them for amplification, but for those who see in them a reflection of the public in which they take part. And they sometimes then become a point of content in and of themselves: what do they measure, what public do they represent, and how should they?

Maybe the question about how algorithms shape culture is the wrong one, or wrong if left by itself. Instead, or at least also, it is about what happens when algorithms get taken up *as* culture, when their particular kinds of claims become legible, meaningful, and contested. We can continue to ask questions about how algorithms shape what is seen or privileged or categorized, about how they may discriminate or make mistakes or treat us like data or automate human judgment. But when algorithms are attending to social and cultural activity, we must remember two things: human activity is public, and algorithmic interventions are too. As Giddens (1984) noted, our scientific attention to human activity is different than to natural phenomena, because people know they are being observed, and act accordingly and often strategically. Algorithmic interventions into human activity face the same challenge. And when algorithmic interventions are also public, in their outputs if not their workings, then they too are observed, taken into account, and strategically contested. This means that the work of algorithms is cultural, and algorithms are not free of culture themselves.

It may be that, because algorithms were so invisible to common practice for so long, it has taken them time to become objects of culture. As Google became prominent, cultural meanings about what it meant to use Google, what it meant to love Google, whether Google was objective or not, and so forth began to emerge and become cultural discourse (Roberge and Melançon 2015). It became a point of reference in casual conversations, the butt of jokes, a verb. But that was a focus on the site, or the service, or maybe the company, not the algorithms specifically. Similarly, Facebook became a cultural reference point as it became more prominent and more widely used, the kind of go-to reference that can be thrown in conversation about anything from life online, to what college kids are like, to an easy dismissal of hyperactive teenagers. In some ways we always domesticate (Silverstone 1994) technologies by pulling them into culture as well as into practice, adding meanings that have, at best, a partial connection to their workings or their purposes. We tame them and make them our own through talk and shared meanings.

So it is as algorithms have become more visible, both as the core functionality of social media and as points of contention in recent controversies, that they too, specifically, could become culturally meaningful. Not only do we begin to discuss what results Google returns or what updates Facebook seems to privilege, but the Google algorithm and Facebook newsfeed algorithm themselves become meaningful, or their algorithm-ness becomes meaningful. When CNN tells us what's trending on Twitter, that's making the algorithm culturally meaningful. When people joke about what the Facebook algorithm must think of them, that's making the algorithm culturally meaningful. When a group of Twitter users try to make their hashtag trend, or celebrate the fact that it is trending, or denounce the fact that it failed to trend, that's making the algorithm culturally meaningful. But, this should extend to algorithms that may not be visible to everyone: stockbrokers find meaning in the algorithms they use, or feel used by; real estate agents have opinions and ideas about the complex systems that now organize their knowledge of the field; police officers tell tales of the predictive analytics that now change the way they work. There is always culture amid the artifice: not just in its design, but in its embrace.

This leaves us with a promising epilogue. Many have expressed concern that users are ignorant of algorithms and their implications, too often treating social media or complex technical systems as either hopelessly inscrutable or unproblematically transparent. Calls for data literacy and concerns about abuses perpetrated by information systems all harbor a fear that users are not concerned enough about the algorithmic systems around them. I think this may underestimate the kind of inarticulate hesitations many users in fact do feel, as well as the outrage around specific cases. But, in the lesson of *American Top 40* and Trends, I think there is a hopeful response to this fear. Users will be concerned about the politics of algorithms, not in the abstract, but when they see themselves and their knowledge, culture, and community reflected back to them in particular ways, and those representations themselves become points of contention. *American Top 40* and the *Billboard* charts did obscure specific biases and underlying ideological assumptions. But they were not embraced unquestioningly. While they reported preferences, they sometimes became objects of contention about those preferences. While they claimed impartiality, they were sometimes challenged for their assumptions and oversights. When they began to seem mismatched with shifting interests in music culture, they were called to task for failing to identify something vital. Their articulation of the hits, or Twitter's identification of Trends, opened up discussions about other trends, other publics, and other possibilities.

Notes

1 The most lucid explanation of the calculations that go into a trending algorithm is from Instagram; it is a very useful primer, as I will not go into much technical detail in this essay: http://instagram-engineering.tumblr.com/post/122961624217/trending-at-instagram (accessed May 26, 2015).

2 Ben Parr, "Twitter Improves Trending Topic Algorithm: Bye Bye, Bieber!" *Mashable*, May 14, 2010. http://mashable.com/2010/05/14/twitter-improves-trending-topic-algorithm-bye-bye-bieber/ (accessed May 26, 2015).

3 Jeff Raines, "Twitter Trends Should Face the Threat of Censorship." *Arts.Mic* August 22, 2011. http://mic.com/articles/1420/twitter-trends-should-face-the-threat-of-censorship#. cjD4342pZ (accessed May 26, 2015); Jolie O'Dell, "Twitter Censoring Trending Topics? Isn't It About Time?" *ReadWrite*, June 18, 2009. www.readwriteweb.com/archives/twitter_censoring_trending_topics.php (accessed May 26, 2015).

4 Avi Rappoport, "Amazonfail: How Metadata and Sex Broke the Amazon Book Search." *Information Today, Inc.*, April 20, 2009. http://newsbreaks.infotoday.com/NewsBreaks/Amazonfail-How-Metadata-and-Sex-Broke-the-Amazon-Book-Search-53507.asp (accessed May 26, 2015).

5 Jon Pareles, "Host in a Big-Tent Era of Pop Music." *New York Times*, June 15, 2014. www.nytimes.com/2014/06/16/arts/music/remembering-casey-kasem-dj-for-a-more-eclectic-pop-radio.html (accessed May 26, 2015); Scott Timberg, "Casey Kasem, Ronald Reagan and Music's 1 percent: Artificial 'Popularity' Is Not Democracy." *Salon*, June 22, 2014. www.salon.com/2014/06/22/casey_kasem_ronald_reagan_and_musics_percent_artificial_popularity_is_not_democracy/ (accessed May 26, 2015).

6 Farhad Manjoo, "How Black People Use Twitter." *Slate*, August 10, 2010. http://primary.slate.com/articles/technology/technology/2010/08/how_black_people_use_twitter.html (accessed May 26, 2015); Lynne D. Johnson, "Reading Responses to How Black People Use Twitter." August 14, 2010. www.lynnedjohnson.com/diary/reading_responses_to_how_black_people_use_twitter/index.html#comment-68768426 (accessed May 26, 2015).

References

Anderson, C. W. 2011. "Deliberative, Agonistic, and Algorithmic Audiences: Journalism's Vision of Its Public in an Age of Audience Transparency." *International Journal of Communication* 5: 529–547.

Ang, I. 1991. *Desperately Seeking the Audience*. London: Routledge.

Baym, N. 2013. Data Not Seen: The Uses and Shortcomings of Social Media Metrics. *First Monday* 18 (10). doi: http://dx.doi.org/10.5210/fm.v18i10.4873.

Beer, D. 2015. "Productive Measures: Culture and Measurement in the Context of Everyday Neoliberalism." *Big Data & Society*, 2(1). doi: 10.1177/2053951715578951.

Beer, D. and Burrows, R. 2013. "Popular Culture, Digital Archives and the New Social Life of Data." *Theory, Culture & Society* 30 (4): 47–71.

Beniger, J. 1989. *The Control Revolution: Technological and Economic Origins of the Information Society*. Cambridge, MA: Harvard University Press.

Beniger, J. R. 1992. "The Impact of Polling on Public Opinion: Reconciling Foucault, Habermas, and Bourdieu." *International Journal of Public Opinion Research* 4 (3): 204–219.

Bermejo, F. 2009. "Audience Manufacture in Historical Perspective: From Broadcasting to Google." *New Media & Society* 11 (1–2): 133–154.

Bimber, B. 1994. "Three Faces of Technological Determinism." In *Does Technology Drive History?* Edited by M. Roe Smith and L. Marx, 79–100. Cambridge, MA: MIT Press.

Bourdieu, P. 1993. *The Field of Cultural Production: Essays on Art and Literature*. Edited by Randal Johnson. New York: Columbia University Press.

Braun, J. 2015a. "Social Media and Distribution Studies." *Social Media + Society* 1 (1). doi: 10.1177/2056305115580483.

Braun, J. 2015b. *This Program Is Brought to You By…: Distributing Television News Online*. New Haven, CT: Yale University Press.

Bucher, T. 2012. "A Technicity of Attention: How Software 'Makes Sense'." *Culture Machine* 13: 1–13.

Burgess, J. and Green, J. 2013. *YouTube: Online Video and Participatory Culture*. Chichester: John Wiley & Sons.

Downey, G. 2014. "Making Media Work: Time, Space, Identity, and Labor in the Analysis of Information and Communication Infrastructures." In *Media Technologies: Essays on Communication, Materiality, and Society*. Edited by T. Gillespie, P. Boczkowski, and K. Foot, 141–165. Cambridge, MA: MIT Press.

Garnham, N. 1979. "Contribution to a Political Economy of Mass-Communication." *Media, Culture & Society* 1 (2): 123–146.

Gerlitz, C. and Lury, C. 2014. "Social Media and Self-evaluating Assemblages: On Numbers, Orderings and Values." *Distinktion: Scandinavian Journal of Social Theory* 15 (2): 174–188.

Giddens, A. 1984. *The Constitution of Society: Outline of the Theory of Structuration*. Berkeley, CA: University of California Press.

Gillespie, T. 2010. "The Politics of 'Platforms'." *New Media & Society* 12 (3): 347–364.

Gillespie, T. 2012. "Can an Algorithm Be Wrong?" *Limn* 2. Retrieved from http://limn.it/can-an-algorithm-be-wrong/ (accessed May 26, 2015).

Gillespie, T. 2014. "The Relevance of Algorithms." In *Media Technologies: Essays on Communication, Materiality, and Society*. Edited by T. Gillespie, P. Boczkowski, and K. Foot, 167–194. Cambridge, MA: MIT Press.

Gillespie, T. 2016. "Algorithm." In *Digital Keywords: A Vocabulary of Information Society and Culture*. Edited by B. Peters, 18–30. Princeton, NJ: Princeton University Press.

Gillespie, T. and Seaver, N. 2015. "Critical Algorithm Studies: A Reading List." *Social Media Collective*. Retrieved from http://socialmediacollective.org/reading-lists/critical-algorithm-studies/ (accessed May 26, 2015).

Gillespie, T., Boczkowski, P., and Foot, K. 2014. "Introduction." In *Media Technologies: Essays on Communication, Materiality, and Society*. Edited by T. Gillespie, P. Boczkowski, and K. Foot, 1–17. Cambridge, MA: MIT Press.

Granka, L. A. 2010. "The Politics of Search: A Decade Retrospective." *The Information Society* 26: 364–374.

Grosser, B. 2014. "What Do Metrics Want? How Quantification Prescribes Social Interaction on Facebook." *Computational Culture* 4. Retrieved from http://computational-culture.net/article/what-do-metrics-want (accessed May 26, 2015).

Halavais, A. 2008. *Search Engine Society*. Cambridge, UK and Malden, MA: Polity.

Hallin, D. 1992. "Sound Bite News: Television Coverage of Elections, 1968–1988." *Journal of Communication* 42 (2): 5–24.

Hallinan, B. and Striphas, T. 2014. "Recommended for You: The Netflix Prize and the Production of Algorithmic Culture." *New Media & Society*. Published online before print June 23, 2014, doi: 10.1177/1461444814538646.

Hanrahan, N. W. 2013. "If the People Like It, It Must Be Good: Criticism, Democracy and the Culture of Consensus." *Cultural Sociology* 7 (1): 73–85.

Havens, T., Lotz, A. D., and Tinic, S. 2009. "Critical Media Industry Studies: A Research Approach." *Communication, Culture & Critique* 2 (2): 234–253.

Helgesson, C. F. and Muniesa, F. 2013. "For What It's Worth: An Introduction to Valuation Studies." *Valuation Studies* 1 (1): 1–10.

Herbst, S. 2001. "Public Opinion Infrastructures: Meanings, Measures, Media." *Political Communication* 18 (4): 451–464.

Hesmondhalgh, D. 2006. "Bourdieu, the Media and Cultural Production." *Media, Culture & Society* 28 (2): 211–231.

Holt, J. and Perren, A., eds. 2009. *Media Industries: History, Theory, and Method*. Malden, MA: Wiley-Blackwell.

Igo, S. 2008. *The Averaged American: Surveys, Citizens, and the Making of a Mass Public*. Cambridge, MA: Harvard University Press.

Introna, L. D. and Nissenbaum, H. 2000. "Shaping the Web: Why the Politics of Search Engines Matters." *The Information Society* 16 (3): 169–185.

Jenkins, H. 2006. *Convergence Culture: Where Old and New Media Collide*. New York: New York University Press.

Lotan, G. 2015. "Apple, Apps and Algorithmic Glitches: A Data Analysis of iTunes' Top Chart Algorithm." *Medium*, January 16. Retrieved form https://medium.com/message/apple-apps-and-algorithmic-glitches-f7bc8dd2cda6#.yzvzfiwcb (accessed May 26, 2015).

McChesney, R. W. 2015. *Rich Media, Poor Democracy: Communication Politics in Dubious Times*. New York: The New Press.

Maguire, J. S. and Matthews, J. 2012. "Are We All Cultural Intermediaries Now? An Introduction to Cultural Intermediaries in Context." *European Journal of Cultural Studies* 15 (5): 551–562.

Mansell, R. 2004. "Political Economy, Power and New Media." *New Media & Society* 6 (1): 96–105.

Marwick, A. E. 2015. *Status Update: Celebrity, Publicity, and Branding in the Social Media Age*. New Haven, CT: Yale University Press.

Morris, J. W. 2015. "Curation by Code: Infomediaries and the Data Mining of Taste." *European Journal of Cultural Studies* 18 (4–5): 446–463.

Mukerji, C. and Schudson, M., eds. 1991. "Introduction: Rethinking Popular Culture." In *Rethinking Popular Culture: Contemporary Perspectives in Cultural Studies*. Edited by C. Mukerji and M. Schudson, 1–62. Berkeley, CA: University of California Press.

Napoli, P. 2003. *Audience Economics: Media Institutions and the Audience Marketplace*. New York: Columbia University Press.

Napoli, P. 2015. "Social Media and the Public Interest: Governance of News Platforms in the Realm of Individual and Algorithmic Gatekeepers." *Telecommunications Policy* 39 (9): 751–760.

Neff, G. 2012. *Venture Labor: Work and the Burden of Risk in Innovative Industries*. Cambridge, MA: MIT Press.

Pariser, E. 2012. *The Filter Bubble: How the New Personalized Web Is Changing What We Read and How We Think*. New York: Penguin Books.

Peterson, R. A. and Anand, N. 2004. "The Production of Culture Perspective." *Annual Review of Sociology* 30: 311–334. doi: 10.1146/annurev.soc.30.012703.110557.

Powers, D. 2015. "First! Cultural Circulation in the Age of Recursivity." *New Media & Society*. Published online before print August 18, 2015, doi: 10.1177/1461444 815600280.

Roberge, J. and Melançon, L. 2015. "Being the King Kong of Algorithmic Culture Is a Tough Job After All: Google's Regimes of Justification and the Meanings of Glass." *Convergence: The International Journal of Research into New Media Technologies*. Published online before print July 2, 2015, doi: 10.1177/1354856515592506.

Rudder, C. 2014. *Dataclysm: Love, Sex, Race, and Identity—What Our Online Lives Tell Us about Our Offline Selves*. New York: Broadway Books.

Salganik, M. J. and Watts, D. J. 2008. "Leading the Herd Astray: An Experimental Study of Self-fulfilling Prophecies in an Artificial Cultural Market." *Social Psychology Quarterly* 71 (4): 338–355.

Sandvig, C. 2015. "The Social Industry." *Social Media + Society* 1 (1). doi: 10.1177/ 2056305115582047.

Silverstone, R. 1994. *Television and Everyday Life*. London and New York: Routledge.

Smith, M. R. and Marx, L. 1994. *Does Technology Drive History?: The Dilemma of Technological Determinism*. Cambridge, MA: MIT Press.

Sorensen, A. T. 2007. "Bestseller Lists and Product Variety." *The Journal of Industrial Economics* 55 (4): 715–738.

Stalder, F. and Mayer, C. 2009. "The Second Index: Search Engines, Personalization and Surveillance." Retrieved from http://felix.openflows.com/node/113 (accessed May 26, 2015).

Sterne, J. 2014. "What Do We Want? Materiality! When Do We Want It? Now!" In *Media Technologies: Essays on Communication, Materiality, and Society*. Edited by T. Gillespie, P. Boczkowski, and K. Foot, 119–128. Cambridge, MA: MIT Press.

Stoddard, G. 2015. "Popularity Dynamics and Intrinsic Quality on Reddit and Hacker News." *ICWSM 2015*. Pre-print retrieved from http://arxiv.org/abs/1501.07860 (accessed May 26, 2015).

Striphas, T. 2015. "Algorithmic Culture." *European Journal of Cultural Studies* 18 (4–5): 395–412.

Tufekci, Z. 2015. "Algorithmic Harms beyond Facebook and Google: Emergent Challenges of Computational Agency." *Journal on Telecommunications and High-Tech Law* 13 (2): 203–218.

Urban, G. 2001. *Metaculture: How Culture Moves through the World*. Minneapolis, MN: University of Minnesota Press.

Vaidhyanathan, S. 2012. *The Googlization of Everything (And Why We Should Worry)*. Berkeley, CA: University of California Press.

van Dijck, J. 2013. *The Culture of Connectivity: A Critical History of Social Media*. Oxford : Oxford University Press.

Weisbard, E. 2014. *Top 40 Democracy: The Rival Mainstreams of American Music*. Chicago, IL and London: University of Chicago Press.

Weltevrede, E., Helmond, A., and Gerlitz, C. 2014. "The Politics of Real-time: A Device Perspective on Social Media Platforms and Search Engines." *Theory, Culture & Society* 31 (6): 125–150.

Williams, R. 1958. "Culture Is Ordinary." In *Conviction*. Edited by N. Mackenzie. London: MacGibbon and Gee.

Wyatt, S. 2008. "Technological Determinism Is Dead; Long Live Technological Determinism." In *The Handbook of Science and Technology Studies* (Third Edition). Edited by E. Hackett, O. Amsterdamska, M. Lynch, and J. Wajcman, 165–180. Cambridge, MA: MIT Press.

Ziewitz, M. 2015. "Governing Algorithms: Myth, Mess, and Methods." *Science, Technology & Human Values*. Published online before print September 30, 2015, doi: 10.1177/0162243915608948.

Zimmer, M. 2008. "The Externalities of Search 2.0: The Emerging Privacy Threats When the Drive for the Perfect Search Engine Meets Web 2.0." *First Monday* 13 (3). doi: 10.5210/fm.v13i3.2136.

4 Shaping consumers' online voices

Algorithmic apparatus or evaluation culture?

Jean-Samuel Beuscart and Kevin Mellet

Introduction

The rapid development of Internet applications and devices has greatly reduced the costs of coordinating and participating in many social and cultural activities. Over the last 15 years or so, there has emerged, through both corporate or individual initiatives, numerous large collectives producing information available to all. Beyond the paradigmatic example of Wikipedia, online video platforms, blog networks, and consumer reviews sites have together built rich data resources, based on free contributions and organized by site administrators and algorithms. These web-based platforms gather heterogeneous contributions from users, which are reconfigured through the operations of selection and aggregation, then sorted and shaped in order to make it meaningful information for their audience. Several terms have been used to describe this mechanism: "collective intelligence" (Surowiecki 2005), "wealth of networks" (Benkler 2006), and "wikinomics" (Tapscott and Williams 2005). The analyses of these authors highlight the ability of such forums to create greater value from scattered individual contributions. They emphasize the efficiency of algorithms and the coordination of technical systems that enable the aggregation of subjective and local contributions into a larger whole that is relevant for users. Overall, these systems and the mathematical formulas that support them, whether simple or complex (based on rankings, averages, recommendations, etc.), are able to build valuable assets from myriad heterogeneous elements produced.

Online consumer reviews (OCRs) are a good illustration of this phenomenon. First popularized by Amazon in the late 1990s, they have since become ubiquitous on the web. They are typically comprised of a combination of a rating (often out of five, and symbolized by stars) and a written review. A product's overall evaluation is summarized by the average rating and the first few lines of some reviews, which the user can freely navigate. OCRs are now present on a variety of sites, particularly those platforms that specialize in collecting opinions (TripAdvisor, Yelp, LaFourchette) and e-commerce sites. They cover a wide variety of goods and services, from hotels and restaurants to funeral homes, as well as books, vacuum cleaners, schools, and everything in between. By bringing together a unified representation of scattered consumer voices, the consumer rating and review system has clearly formed a large part of our collective digital intelligence. Indeed, the creators of these sites themselves often invoke democratic legitimacy by presenting themselves as the voice of ordinary consumers. As with democratic elections, every consumer is allowed one vote, and all opinions are presumed equal. For example, the CEO of TripAdvisor has stated:

> Online travel reviews have hugely changed the way the travellers can plan their holidays—they add an independent view of where to go and stay giving another level of assurance that their hard earned travel Euro is spent wisely. [...] That's the positive power of Internet democracy in action.
>
> (Kaufer 2011)

A further claim to legitimacy is the strong consumer appetite for these services, as a majority of Internet users say they use them regularly; this has translated into tangible effects in many markets. Indeed, several marketing science and econometric studies have demonstrated a significant impact of OCRs on economic activity in sectors such as hotels, restaurants, and cultural consumption (see references below in the section "The uses of ratings and reviews").

While it has received a lot of media commentary, the practice of rating and reviewing has received very little empirical research. The few that exist, mainly in sociology and organization studies, are schematically divided into two categories. The first investigates the motivations of those who frequently contribute comprehensive reviews, emphasizing the importance of recognition, skill development, and gratification: according to these studies, OCRs appear primarily to be the work of semi-professional evaluators (Pinch and Kessler 2010), somewhat leaving ordinary contributors on the margins. A second category insists instead on the heterogeneity of scattered, subjective contributions, stressing the decisive role played by algorithms in constructing meaningful assessments, overall scores, and rankings (Orlikowski and Scott 2014). These analyses support the perspective of broader reflections on collective intelligence, highlighting the crucial role of algorithmic synthesis, and calculations more generally, in the aggregation of subjectivities; they suggest that contributors are largely isolated, guided by an irreducible subjectivity, and, statistically speaking, independent.

Recently, web-based platforms such as OCR websites have gained the attention of scholars for their capacity to organize information and make sense of users' contributions. By aggregating and sorting contributions through proprietary and often undisclosed algorithms, these websites have a great ability to shape culture (Striphas 2015). Through their algorithms, they are in a position to redistribute valuations and preferences within many cultural and information industries, in ways that cannot be democratically discussed or disputed (Gillespie 2010; Morris 2015). Though these analyses raise a crucial point—our ability to discuss what's valuable in our cultures—they tend to presume that the effect of the algorithm is complete and undisputed. From a Foucauldian perspective, they stress the power of web platforms to organize users' information, and consider the algorithm as the result of an explicit strategy; conversely, users are mainly seen as passive subjects. In this chapter, we try to qualify this perspective by underlining the role of users in shaping algorithmic valuation. As stated by Oudshoorn and Pinch (2003), "users matter" in the shaping of technologies; in our case their actions shape these platforms in at least two ways. First, they interpret the information provided by the platforms, select and weigh it in a way that is not completely scripted by the site. These interpretations are based on their experience, and they have good reasons to adhere (or not) to the site's valuation standards. Second, users shape the platform through their contributions, by choosing whether or not to follow its guidelines, and by anticipating the effect of their actions. As a consequence, the 'algorithmic' valuation is co-produced by the site and its users through a relationship that cannot be interpreted as pure passivity and alienation. Following MacKenzie (2011, 2014), the set of interpreting schemes and practices developed by users around the website can be called an "evaluation culture."

In this chapter, we follow this user-centered perspective by highlighting the collective practices and reflexivity of ordinary contributors. We show that the authors of such opinions do not give free rein to their subjectivity, but write in consideration of a specific target audience and/or website. There exist common assumptions and norms concerning the proper format and content of an opinion, as well as standards governing what makes a contribution helpful, as well as a rating relevant. All of these standards can be described as part of evaluation culture as described by MacKenzie; as well, the development of a new assessment tool is necessarily accompanied by the emergence of more or less coherent methods of interpretation, reading practices, and the manipulation of instruments. Rather than contributors primarily seeking recognition or consumers governed by their subjectivity, it is the figure of a common user who is reflexive, knowledgeable, and accustomed to these services that we want to highlight here. In order to do this, we rely on a survey of contributors to the restaurant rating website LaFourchette (www.lafourchette.com), supplemented by contextual data from the web, as well as a survey of a representative sample of consumers.

The first part of this chapter presents a brief literature review, centered on empirical findings concerning the use of ratings and reviews. The second part is devoted to the presentation of LaFourchette, and the methodology used in this study. The third part focuses on the motivations of users who contribute to the site, particularly through their practices of reading the ratings and giving advice to consumers: participation is primarily motivated by a satisfactory reading experience, and influenced by a certain understanding of the collective work done by website users. The fourth section describes the standards that form the evaluation culture of the site in terms of form and content, and attempts to sketch in broad terms the figure of the contributor 'socialized' to these types of sites.

Ratings and reviews, their uses and academic research

The uses of ratings and reviews

Consumer reviews are now a standardized tool, ubiquitous on the web and fully integrated into the lives of Internet consumers. The format, introduced by Amazon in the late 1990s, allows users to express themselves through a combination of a rating system and written text (Beauvisage *et al.* 2013). Reviews and ratings are found on most e-commerce sites, and also on websites dedicated specifically to the assessment of goods and services by consumers. In the field of tourism, for example, TripAdvisor collects reviews and opinions on hotels and restaurants around the world, and had approximately 25 million unique monthly visitors in the U.S. in 2015, and the same number in Europe. Such sites exist for a wide variety of markets, such as shops and local services (Yelp, with 83 million unique visitors in the U.S. in 2015), restaurants (Zagat, Urban Spoon, LaFourchette), consumer goods (Ciao), and so forth.

A great deal of converging data demonstrates the increasing incorporation of online reviews and ratings into the everyday consumption practices of Internet users. Our survey among a representative sample of French Internet users shows that 87 percent of them pay attention to reviews, and 89 percent say they are useful; 72 percent of them have contributed an online review or opinion, and 18 percent say they do it often (Beauvisage and Beuscart 2015; see "Methodology for the study of LaFourchette" below). Despite the likely over-reporting bias, the steady growth in the positive response rate to these questions shows the increasing popularity of this practice. Another indication is provided by the effect of ratings, as measured by econometric investigations seeking to evaluate the impact of online reviews on sales: overall positive effects were observed for book sales (Chevalier and Mayzlin 2006), cinema tickets (Liu 2006), and restaurant sales (Luca 2011).

This expansion of review websites can generally be understood in two distinct ways as part of the recent democratization of evaluation (Mellet *et al.* 2014). On one hand, these sites greatly extend the scope of evaluated goods and services, and thus the number and type of consumers who are reached. For example, in the case of restaurants in the French market, the *Michelin Guide* lists about 4,000 restaurants, mostly upscale and classy; for its part, TripAdvisor provides assessments of 32,000 establishments, 60 percent of them with meals available for €30 or less. On the other hand, review websites allow all consumers to offer their opinions, popularizing the process initiated in the late 1970s by guides such as Zagat, which began collecting consumer opinions via written questionnaires. By 2012, for example, TripAdvisor had collected 338,000 reviews of restaurants across France, collected from 178,000 distinct users.

Academic research

Relatively little empirical work has been conducted on the contributors of these sites, or the meanings they ascribe to their assessment activities. The pioneering work of Pinch and Kessler (2010) on the most active Amazon contributors shows that they are mostly male, have an above-average education, and are often engaged in activities related to knowledge-production. In terms of motivation, the collected responses (by questionnaire) highlight several dimensions: personal development opportunities (writing skill, expertise in a certain area), recognition from other members on the site, and the material and symbolic rewards offered by the site. Other, more recent studies have confirmed these findings: King *et al.* (2014), in a survey of marketing studies on this topic, place self-esteem (self-enhancement) and the search for recognition as the strongest motivations for writing OCRs. From the same perspective, Dupuy-Salle (2014) shows that recognition issues are strongly correlated to membership among the elite contributors of film reviews sites. While convergent, however, these results cover only a very small minority of overall OCR contributors.

A second approach common in recent research focuses more on the content of the written reviews and opinions. Beaudouin and Pasquier (2014) observed that online opinions of films vary between two poles: some opinions strive to resemble professional criticism, to construct an objective and argumentative discourse, while others are characterized more by the expression of subjective experience, often written in the first person. Other research is more interested in examining how speakers assert their qualifications when reviewing goods and services. Jurafsky *et al.* (2014) note that reviewers demonstrate their competency in about 25 percent of online reviews for consumer goods (e.g., "I cook a lot of duck," "I'm on my second child," etc.). Other research examines how the quality of goods is evaluated. Cardon (2014), in a textual analysis of opinions on French hotels on TripAdvisor, found a strong focus on certain attributes (e.g., the quality and quantity of breakfast) at the expense of evaluations of more traditional criteria in the industry. Finally, some studies suggest that opinions generally vary according to the type of good being evaluated—they are typically longer for expensive products (Vásquez 2014)—and often depend on the rating given: for hotels, reviews tend to be more fact-based when they are less favorable (Cardon 2014).

Methodology for the study of LaFourchette

In this chapter we rely on interviews with contributors to the site www. lafourchette.com. This qualitative material is supplemented by contextual data from the web, and by the results of a survey of a representative sample of Internet users.[1]

Launched in 2007 in France, LaFourchette is a restaurant review website (and mobile app) characteristic of the second generation of platforms dedicated to local goods and services that appeared between 2000 and 2008 (TripAdvisor, Yelp, Qype, Tellmewhere, etc.). Unlike the online city guides of the first generation, created in the 1990s (e.g., Citysearch in the U.S., Cityvox in France), these newer platforms are characterized by the lack of a strong, central editorial authority, by the participation of Internet users (as both consumers and merchants) to enrich the content and inclusively evaluate places and goods, and by the *a posteriori* algorithmic moderation of the results (Mellet *et al.* 2014). Thus, these websites put the participation of users at the heart of their activity— and of their business model, which, in one way or another, is based on the monetization of content and traffic generated by users. And they try hard to encourage and organize it. First, OCR platforms have developed specific tools in order to foster participation. The most common incentive apparatuses mobilize social features, such as user profile pages, badge systems to reward the most prolific contributors, internal communication tools, etc. These devices tend to single out contributors and give greater weight to the reviews of the most prolific authors (Mellet *et al.* 2014). Second, OCR websites try to shape the contributions in order to make them relevant to the audience and to the industry they address. As market intermediaries, they design tools in order to favor appropriate matches, by encouraging contributors to respect specific formats. Through their forms and input fields, they encourage users to follow existing shared criteria of valuation.

While the presence of certain features on the site, and its acquisition in 2014 by TripAdvisor, strongly root LaFourchette in a typical participatory online model, some uses of the site are quite specific, as we shall see below. LaFourchette is essentially a software platform with an incorporated reservation system used by about 4,200 French restaurants (as measured in July 2012). Users can navigate through the pages of those restaurants and make reservations; they can take advantage of rebates from certain restaurants, who in return receive greater visibility on the platform; and once they have eaten at the restaurant, users are invited by e-mail to give a review and a rating. This invitation to contribute is the principal incentive mechanism we observed, since at the time of the survey contributions were not encouraged by rewards nor elaborate badges—except the inconspicuous 'gourmet' status obtained after the second review, and 'gastronome' after the tenth review. Furthermore, the evaluation form on LaFourchette is similar to that found on other sites. Contributors are first invited to rate three criteria from 1 to 10: food, service, and setting. The individual score given to the restaurant is produced from the (publicly displayed) weighting of the three ratings. Then, contributors are invited to write in a free text field. There are no explicit instructions or recommendations, and no apparent limit: "it is almost unlimited in size," a LaFourchette manager reported to us.

More than 642,000 ratings were posted on the site as of July 2012, an average of 153 per restaurant. While the vast majority of the 292,000 contributors have reviewed only occasionally—87.5 percent have left three reviews or fewer—a significant number of users are more active: 13 percent of contributors have left four reviews or more and account for half of all posted reviews; among them, 2.8 percent of contributors have posted ten reviews or more, and there are about 3,000 users (0.1 percent of all contributors) who have left more than 50. In this chapter, we are most interested in these regular contributors.

Overall, we interviewed 33 people who responded positively to a request sent by LaFourchette to a random sample of 100 users, consisting of 21 very active contributors (with over 50 reviews posted on the site) and 12 somewhat active contributors (10–15 reviews). Most were also contributors or visitors to

TripAdvisor, so they were asked about their use of this website as well. The interviews lasted 30–90 minutes and were conducted face-to-face (24) or online by video link (9) in November 2013. They were transcribed and fully encoded using the QDA Miner qualitative data analysis software. The sample includes 14 women and 19 men, mostly from Paris (14) and the greater Paris region (13). On average they are 48 years old, and have a high level of education—22 have four or more years of university education. They often visit restaurants, and attend on average one or two per week (up to eight per week). They dine at restaurants in a personal context (as a couple, or with family and friends), and one-third of them visit restaurants in a business context. Overall, they have contributed to LaFourchette for over a year and a half, between ten and 194 reviews each.

Contributing to a collective goal

The first key issue for us in interpreting rating and review systems is the meaning given to this activity by the contributors. While our investigation uncovered some of the reasons discussed in previous work on highly active contributors (pleasure, great interest in the subject), other motivations were also found, including the explicit desire to contribute to a collective goal that is considered useful and helpful. We focus first on their experiences as readers, before analyzing the scope of their motivations for contributing.

Experience as readers of online reviews

All the users surveyed expressed having had an excellent experience with LaFourchette. Although this finding may be magnified by selection bias (individuals who enjoy the website are more likely to talk at length with sociologists about it), all users without exception voiced satisfaction with the site. The LaFourchette website (and by extension, TripAdvisor) is seen as a highly reliable tool for choosing a restaurant, regardless of the context or requirements. Those who once used traditional guides abandoned them in favor of online sources; the most gourmet among them continue to consult the *Michelin*, but only for high cuisine. For those who frequent restaurants less often, LaFourchette is their first choice. In general, the guiding idea of the site is that of "discovery without risk": through users' accounts, the site maps a broad range of possibilities, all while minimizing unpleasant surprises:

> I must admit that La Fourchette allowed me to change my address book a little, that is to say, to include addresses ... of restaurants where I would never have gone before.
>
> (E26)

> It's true that this type of application has changed us as consumers ... now I would not eat at just any restaurant at random.
>
> (E11)

These positive experiences are based on a number of reading patterns common among most users. To get an idea of a restaurant, they combine the information available on the site (rating, reviews, photos, menus, etc.), but never limit their impression to just the rating. They read the reviews, at least the ten shown on the first page, and often visit multiple pages. But while they make their choice by combining these criteria, it seems to be strongly influenced by a good average rating as a primary criterion. For example, two-thirds of respondents reported a score below which they would not consider a restaurant: most often 8/10, 7.5 for some, and 8.5 for others. Recall that the average scores are high on these sites: the median rating is 4/5 on TripAdvisor, and 8/10 on LaFourchette; most users generally do not consider restaurants ranked eight or lower, and thus limit their choices to the best rated restaurants. Relying on these typical and common patterns of judgment, most users consider the experience of recommended choice offered by these sites as reliable and rewarding.

Interestingly, this account contrasts sharply with journalistic writings devoted to the topic of online reviews, which focus mainly on the issue of "false information" and fraud.[2] At no time during the course of the interviews did interviewees express distrust with regards to fake reviews. When asked about it, they recognized that some opinions can sometimes seem dubious, but that this never tainted the reliability of their judgment, given that the reviews are relatively convergent and numerous (at least 20 for some users, up to 50 for others). This is especially true for LaFourchette, where reviews are connected to a reservation in the restaurant, and traced by the site; but it is also true, though less unanimous, of TripAdvisor, whose assessments are considered reliable since there are so many contributors. The quantitative survey (Beauvisage and Beuscart 2015) produced a similar result: while 90 percent of Internet users admit to having seen one (or more) fake review(s), 76 percent believe that "this does not prevent them from getting a good idea" of the restaurant's quality. The dominant narrative is thus that LaFourchette and TripAdvisor offer a satisfying experience and highly reliable judgments.

Contributing to the collective

Writing a review can be done relatively quickly, with users on average devoting between five and ten minutes to the activity, usually the same day or the day after their experience. From this perspective, the reminder e-mail prompting them to give their opinion following their meal is an efficient means for getting users to write: several interviewees mentioned it as a reason for contributing.

When discussing their motivations, some contributors (7 out of 33) described the pleasure they take in writing reviews. Several themes emerged: their opinions will extend and deepen the experience of dining out; for those who love to write, choosing the right words is pleasant in itself; and more broadly, it is enjoyable for many to offer advice. For example, here are some excerpts from LaFourchette contributors, which highlight the pleasures of writing and expressing their interest in a cultural field:

"I like it a lot. I really like to share, it makes me happy" (E33); "I really like it, I offer opinions all the time" (E25); "We do it for fun, otherwise we wouldn't do it at all" (E1); "I take great pleasure in reviewing" (E10); "Yes, it's a pleasure" (E15); "It's entertaining" (E20).

Compared to the existing research, however, this aspect of pleasure is relatively minor in our investigation. One of the other dominant motives in the existing literature, the search for recognition, is also completely absent in our study. This is explained in part by LaFourchette's site design, which at the time of the investigation offered no features characteristic of social networking websites: links among 'friends' or 'followers,' comment threads, likes, favorites, and so forth. But it is also based on the aggregated choices of users, who reflect the site's main uses: none of the users surveyed had completed a user profile, uploaded a profile photo, etc.; further navigation throughout the site confirms that completed profiles are exceptions to the rule. Even though they post many reviews, LaFourchette users thus do not contribute in order to increase the visible activity of a profile, or as a source of recognition. Also, when asked about their sense of belonging to a "specific community" of LaFourchette members,[3] most respondents answered in the negative. None of them had any social relationships with other contributors, and most wished not to have them.[4]

> The term "community" is a bit much … I'm glad to be part of the site and enjoy contributing, yes, without reservation. I feel absolutely no pressure. I think there's a real interest in the site, so I'd say I participate gladly. But to say that I'm part of a community … no, I don't really have that impression. That's a bit strong of a term, in fact.
>
> (E2)

Instead, the dominant motivation appears to be an anonymous and meaningful contribution to the public good. Users emphasized their need to maintain an overall reciprocal system: they offer their opinions to contribute to a system from which they benefit. The coding of the interviews revealed a wide variety of expressions of collective participation: "it's part of the game" (E2), "I want to return the favor" (E4), "I want to fulfill my contract" (E10), "it's a give-and-take" (E28, E30), "it's win-win" (E10), "it's only fair" (E26), "it helps" (E12), "it is my duty to inform people" (E27), etc. The primary motivation for writing a review, in our survey, appears to be a feeling of responsibility, a moral obligation to contribute to the collective good, and a refusal to take advantage of the system.

   To clarify the logic underlying users' motivations to contribute, the interviews oriented the discussion towards the target audiences of the posted reviews: in most cases, it is above all other users who were identified. Among the components of the socio-technical assemblage built around these sites— linking together a website, restaurateurs, search algorithms, and other users/ evaluators—it is by far the users who are mentioned as the primary recipients of their contributions, and those to whom they were also indebted: "When I write, it is mostly for consumers" (E3), "for people like me who go to restaurants. Because I think that it can be useful to someone" (E5), "from the moment when I started to enjoy reading peoples' opinions, I also began offering my own" (E32), "I think it's worth it to read the views of others, and I guess mine as well" (E14), etc. To summarize:

> So, first, I'm a reader of reviews, and I think a part of my purchases are based on the advice I received. Since I attach importance to this activity, it also seems important to me that I leave my own opinions.
>
> (E12)

In addition, some users address their opinions directly to the restaurant-owners. On one hand, this is done to thank the establishment and its staff for a good experience: "if I am satisfied, I leave a comment to keep encouraging them in the right direction" (E17). On the other hand, these consumers also feel that their role on review sites is to help restaurant owners—possibly in spite of their efforts—through criticism, which they insist is always "constructive." To contribute is thus also to participate in improving the restaurant experience, in addition to guiding consumers:

> I always take care to comment with a constructive purpose in mind. I am not a mean or abusive critic who contributes nothing. With constructive criticism, I feel I can help improve the service.
>
> (E15)

> For them, I think it's important, because it can make things better, or it can help show them that there are some good qualities, or flaws too. So by rating and reviewing them, I think you can perhaps help them be aware of and remedy the problems.
>
> (E25)

This analysis of the motivations for contributing thus outlines a discrepancy with the literature on online participation. Rather than contributors participating in a group in order to refine their skills, gain recognition, or receive material and symbolic rewards, our survey suggests that they are satisfied with their anonymity—none claimed to take any steps that would allow them to be recognized— and participate primarily in order to contribute to the collective good. They write in anticipation that their reviews will be read by other members of a sociotechnical collective that they themselves deem useful. These users are thus more self-reflexive and moral than assumed by much of the academic literature, and especially by the media.

A shared definition of a 'good review'

The second key element of this investigation, further supporting the figure sketched above of a reflexive user contributing to the group, is the widely shared definition of what constitutes a good review. This definition outlines the contours of an 'evaluation culture' common to regular users of the site, which can be understood as a set of representations and practices surrounding the best way to read and participate in the collective.

The proper format

The majority of users agreed on the fact that the best review is a short review. Contrary to the image of egocentric individuals recounting their personal experience in detail, the proper length of a review was estimated to be between three and five lines: it must "cut to the chase" (E1), "be synthetic" (E25) and "summarize" (E26). "Four or five lines maximum, it's not a novel," said E9. Some regular contributors to the site believe that their opinions have shortened over time and as they use the site more often. Longer explanations are justified only if they emphasize a point particularly relevant to the establishment, whether positive or negative. It is justified when "highlighting some thing that you really liked, such as an item on the menu" (E1), or a contextual element, such as "the bathroom was completely vintage, it was really extraordinary"; likewise, many suggest expanding on a review "when the experience was very bad" (E30). Even in these cases, elaborations should take only one or two additional lines: "If everyone starts to write ten lines, it's over" (E27).

As previously indicated, the standard format of reviews is based on users' previous experience reading them, which they feel are typical of those reading their own opinions. When seeking restaurants, users generally browse 10–20 opinions—in addition to accessing other available information—reading quickly, diagonally, seeking to identify similarities, patterns, and salient features. Respondents stated that it is best when there are numerous opinions, especially those that directly match their criteria; in addition, they will often isolate negative opinions to assess their significance within the overall pattern. Generally speaking, since there are usually fewer of them, negative opinions are considered to be related to specific situations or atypical customers ("grumpy customers, I don't pay much attention to them," said E10), unless the criticism concerns hygiene:

> I try to look at two or three bad reviews and ask: "Okay, what's happening here? Is this an isolated occurrence? Was the server cranky and thus poorly reviewed?" Off days can happen sometimes, and then everything goes wrong in the kitchen.
>
> (E13)

> If I see something that keeps coming up in restaurant reviews that concerns hygiene or cleanliness, then it's a no-go.
>
> (E1)

We can thus define the best review format based on contributors' reading practices. Reviews are intended to be quickly scanned to confirm an emerging evaluation or add a new element to it. What is sought is thus not the subjective evaluation of a specific consumer, but a contribution to an evaluation formed from previously read reviews: either a confirmation of a salient point, or a critical nuance. In this context, "it's annoying to see reviews that are ten kilometers long" (E11): the best reviews are short, get to the point, and do not go on too long when they add an original element to the evaluation.

Note that some interviewees (four out of 33 in our sample) significantly deviate from this predominant standard. They recognize that "sometimes reviews are a bit long but it does not bother me, as long as there's space" (E21), they claim to "write essays" (E28). Two of them in particular are users with a strong relationship with writing: one is a writer, and the other "is known in her family for her incisive style" (E21). These users may circumvent the conventions of the format because of the high value they place on their writing; or perhaps their attachment to creativity predisposes them more to think of the website as a forum for subjective expression, while most users reject this vision and those practices associated with it.

Evaluation criteria

Users also strongly agree on what constitute good evaluation criteria, which comprise a second key element of the culture built around review sites. As we noted above, these criteria are strongly guided here by LaFourchette, which invites reviewers to separate their scores into three main components, which are then aggregated into a total score: food, setting, and service, optionally complemented by an appreciation of the value for money. When questioned on the criteria they usually assess in their written opinions, users spontaneously mentioned these three dimensions, which they consider an appropriate and meaningful way to account for the restaurant experience.

> I speak of three points on LaFourchette. The food is what counts above all, followed by service and value.
>
> (E3)

> Yes, yes, these are the elements that interest me in a restaurant: the setting, reception, kitchen, and service. These are four elements that I systematically give an opinion about, almost exclusively.
>
> (E16)

> I try to address reception, price, and quality of food in a systematic way.
>
> (E17)

Here, the prescriptive role of the platform appears quite explicit. Users are clearly guided by the strategic choices of the site's managers, embedded in the user interface. That said, this framing is perceived as such, fully accepted and even endorsed by users, who use it as a prominent and conventional cognitive marker to write their assessments when they could just as easily express their free-form subjectivity in the open field.

With regards to the quality of the food, assessments are typically simple and conventional. Consistent with the length requirements, accounts of the meal do not take the form of food criticism, descriptions of flavors, or subjective feelings of taste. Rather, they tend to simply verify that the food lives up to its promises, based on simple, widely shared criteria: the freshness and quality of the food and the absence of major technical faults (in preparation or seasoning, for example), with possible mentions of portion size, taste, and overall satisfaction.

> For me, the first criterion will be the quality of what I eat. Not the quantity, but quality. As I mentioned, when I go to a restaurant, I expect the food to be fresh.
>
> (E5)

> We always insist on mentioning when there is good food with fresh produce and well-prepared meals.
>
> (E32)

Restaurant review sites, and LaFourchette in particular, are not devoted to amateur gastronomic criticism, at least in the sense of offering elaborate, subjective accounts of unique aesthetic experiences. Among our respondents, the great majority make no claims for their qualifications or skills; and those who do claim to have gastronomic experience believe that these sites are not the place for such criticism: "I do not blog," states one respondent (E1). Reviews, however, are largely centered on the food, though they offer them in a more general, less subjective way. Evaluations of the freshness of the food, cooking methods, and portion size do, of course, require some skill and are subject to individual variations, but far less than subjective assessments of flavor combinations, for example. Again, the effectiveness of review sites in general is based not on the relevance of highly refined evaluations, but rather on the accumulation of conventional, converging assessments.

The most active contributor in our sample (281 ratings and 194 reviews) exhibits a deep familiarity with gastronomic culture and much experience in the foodie world. An avid reader of gourmet guides and blogs, he displays a virtuosic ability to describe food and restaurants. However, on LaFourchette he fully adopts the conventional assessment standard of brief reviews:

> Rather than go into all the details, to say, "yes, I have eaten such-and-such a dish which was excellent. By contrast, this other one was very bad…" this doesn't do much. I think we should be a little more concise. […] In the beginning, as I mentioned, I was perhaps a little more expansive in my opinions. I used to write maybe five or six lines, and sometimes, when

exaggerating, up to ten. Today I stick to about three lines. That is to say, over time, and as I read more and more comments ... [...] I do not do analysis, [I'm not] like Gilles Pudlowski, or François Simon.[5]

(E30)

In addition to the aforementioned evaluation criteria (food, service, setting), which are prescribed by the platform, contributors often add another one, to which they attach great importance: "reception" (i.e., how they as customers are greeted). This term is found in 26 (of 33) interviews, without being referred to by the site itself, nor proposed by the interviewers. This suggests it is an essential quality for restaurant reviewers, and probably more specific to amateur assessments. Separate from service, reception means "the friendliness of the people" (E6), "people who smile, and those who naturally want to please you" (E5); as summarized well by E28, good reception is "when, as soon as you enter, you feel welcomed as if you're friends of friends." Conversely, poor reception is too formal, instrumental, and a little cold.

> The welcome in particular is very important.... It's the first impression you get of the restaurant. If we arrive and they're cold, there's no "hello," not even "Do you have a reservation?" it's just "sit down over there!"... When it's not warm, we may even leave discreetly. For me, this "hello" is very important, the smile is very important.

(E33)

In summary, when selecting evaluation criteria, contributors are guided to a certain extent by the criteria put forward by the website. Indeed, they base their assessments on traditional criteria—service, food, setting—that guide professional evaluations (Blank 2007) and are included in the scoring criteria of the website. However, these amateur reviews clearly stand out from professional assessments in general, and from the explicit prescriptions of LaFourchette in particular, in several ways: the evaluation of the food remains relatively procedural, focusing on quality and the absence of major technical errors; and the description of service is coupled with an assessment of the reception, i.e., the ability of a restaurant to put ordinary customers at ease and treat them kindly, as expected.

Evaluation culture

Contributors to LaFourchette, and to a lesser extent TripAdvisor, share a consistent set of practices of reading and writing online reviews. They offer short, summary opinions capable of being quickly and easily read and understood by many others. Assessments are based on shared criteria that are suggested by the site and taken into account by users; they typically also consider the reception, a dimension more specific to amateur evaluations. This criticism is explicitly not based on refined tastes, nor on virtuosic gastronomic experience; as often as not, it is procedural, verifying that restaurants meet the basic expectations in terms of quality of food, service, etc. This is consistent with the practice of reading online reviews, which tend to build an evaluation of a restaurant by weighing the accumulation of a large number of opinions, discounting or dismissing those that are too subjective—both positive and negative—to arrive at cumulative appraisal of key criteria. These judgments are thus socialized, in the sense that they are somewhat determined by their anticipated use by other consumers; they meet a set of conventions considered to be collectively relevant.

This description is far removed from the common notion of isolated consumers freely expressing subjective feelings (whether joy or frustration) on the site, which then derives meaning from the mere accumulation of these disjointed voices. Existing descriptions often insist on the unregulated nature of online evaluations:

> Valuations—which have traditionally been produced by a small number of recognized experts, critics, media, and authorities using formal, standardized and often institutionalized criteria grounded in professional knowledge and industry experience—are now also (and increasingly) being produced online by large numbers of anonymous and distributed consumers using informal, variable, and individual criteria grounded in personal opinions and experiences.
>
> (Orlikowski and Scott 2014, 868)

Amateur assessments are described by Orlikowski and Scott as based on unstable and personal criteria—"personalized and often contradictory assessments," "volatile assessments of a distributed and disembodied crowd"—and taking erratic formats: "Reviews vary in length from a sentence to a short essay," and appear in "various styles." For these scholars, it is ultimately only the site and the algorithm that deserves credit for producing meaning: TripAdvisor's ranking algorithm "expresses the unregulated and anonymous opinions of many consumers."

On the contrary, we suggest that at least part of the effectiveness of this phenomenon is the ability of users to build a coherent pattern of use that regulates their evaluation behavior to work towards a collective aim. The site is built around an "evaluation culture" (MacKenzie 2011), which guides users on how to read and write evaluations. MacKenzie has identified several criteria for qualifying a set of cultural representations that guide evaluations, which are partially met here. In particular, the user evaluations share a common 'ontology,' or a definition of what fundamentally gives value to a restaurant. As well, contributors go through a 'socialization' process within the socio-technical system, and they learn to recognize and replicate the best practices: reviews are reduced in length over time, are more to the point, and give only the most useful information. However, this occurs only indirectly, through the imitation of other users, since there is no direct interaction among the site's members.

This socialization into evaluation culture is of course uneven across individuals. We noted in our sample some minor deviations regarding the format or the intended audience of the reviews. Above all, infrequent contributors (those with one or two reviews, excluded from our sample) do not generally adhere to the dominant uses of the site. An exploratory survey of these contributors suggests that they are primarily guided by a desire to express a strong sense of satisfaction or anger, which brings them more in line with the portrait of typical contributors described in the existing literature (Pharabod 2015). Overall, through repeated use of the site, contributors realize and integrate conventions of evaluation culture.

Review sites ultimately rely on the balance between two components. On one hand, a minority of regular users is familiar with the culture of evaluation as we have described it; on LaFourchette, users who have posted five or more reviews represent 13 percent of contributors and are responsible for just over half of all opinions. On the other hand, a majority of occasional visitors exhibit less consistent contributions, which are more like the "unregulated opinions" mentioned in the literature: in our case, 66 percent of contributors have written only one review. It seems that through their practice, users become accustomed to the standards and good practices of decentralized evaluation, learning to control the expression of their subjective opinions for the collective good.

Conclusion

Our research allows for an enriched understanding of how distributed evaluation sites function, as they grow in importance across many industries. Besides the two ideal contributor types previously identified by the literature—intensive participation of an 'elite' group driven by the quest for recognition, and the isolated expression of 'unregulated opinions'—we highlight a third: regular contributors who are part of a coherent evaluation culture, shaping their participation according to a collective aim. These users (approximately 10–15 percent of all contributors) comprise the heart of review sites, not only because they produce a majority of the evaluations, but also because they maintain standards and good practices, and habituate new contributors through their example.

This analysis also allows us to comprehend the operation of collective intelligence produced by this type of platform. The production of meaning and intelligence is not only based on the aggregation algorithm, on the ability of formulas and site design to collect disparate contributions by extracting their unique and singular meanings. Rather, much of this work is actually conducted by the users themselves, through reflexive feedback loops between their reading and writing practices, deduced from the good practices that are inherent in this shared culture. Contributors to these sites assume the codified role of evaluator and adjust their contributions accordingly. Though their participation is partly prescribed by the site, it is perceived and accepted as such, because the framing is viewed as relevant for the readers. This participation also overflows the framing, by adding specific qualities such as "warm reception." In this sense, though users' contributions generate economic value for the platform (Scholz 2013), writing a review is not considered as 'labor' by users, but as a contribution to a system they find very useful. The algorithmic culture of these sites is thus both a guide to reading and interpreting the reviews and rankings they produce, and a set of practices that contribute to the overall effectiveness of the evaluation process.

Acknowledgments

We would like to thank LaFourchette, who supported the survey on which this article is based. Thomas Beauvisage, Marie Trespeuch, and Anne-Sylvie Pharabod helped out with several aspects of this investigation, and their work and comments were invaluable to the development of this article.

Notes

1 The usage data for www.lafourchette.com and www.tripadvisor.fr (number of restaurants listed, number of contributors and reviews) were extracted using ad hoc tools in July 2012 (cf. Mellet *et al.* 2014). The extraction and processing of the data was conducted by Thomas Beauvisage. We also rely on a questionnaire survey conducted by Orange Labs and Médiamétrie in November 2013 of a representative sample of French Internet users (n=2,500). This quantitative survey focused on both consulting and writing online reviews and ratings (Beauvisage and Beuscart 2015).
2 For an exploration of cheating on review sites, see Reagle (2015).
3 The term "community" is systematically used by managers of review sites, and LaFourchette is no exception. References to "the LaFourchette community" are everywhere on the site. Note, however, that some platforms, such as Yelp or TripAdvisor, have implemented active strategies to build and manage visible social interactions among contributors: customizable profile pages, badges, communication tools internal to the site, the organization of events in physical locations, etc. (Mellet *et al.* 2014).
4 One interviewee proved to be an exception: an intensive user of both TripAdvisor and LaFourchette, he is very attentive to the management of his profile on the latter site, and does not hesitate to use it in his negotiations with restaurants and hotels.
5 Pudlowski and Simon are two of the best-known food critics in France.

References

Beaudouin, V. and Pasquier, D. 2014. "Organisation et hiérarchisation des mondes de la critique amateur cinéphile." *Réseaux* 32 (183): 123–159.
Beauvisage, T. and Beuscart, J.-S. 2015. "L'avis des autres." digital-humanities.orange. com, September 2015. Retrieved from http://digital-humanities.orange.com/fr/ publications/dossiers/articles/595-lavis_des_autres (accessed November 20, 2015).
Beauvisage, T., Beuscart, J.-S., Cardon, V., Mellet, K., and Trespeuch, M. 2013. "Notes et avis des consommateurs sur le web. Les marchés à l'épreuve de l'évaluation profane." *Réseaux* 177: 131–161.
Benkler, Y. 2006. *The Wealth of Networks: How Social Production Transforms Markets and Freedom*. New Haven, CT and London: Yale University Press.
Blank, G. 2007. *Critics, Ratings, and Society: The Sociology of Reviews*. Lanham, MD: Rowman & Littlefield.
Cardon, V. 2014. "Des chiffres et des lettres. Evaluation, expression du jugement de qualité et hiérarchies sur le marché de l'hôtellerie." *Réseaux* 143: 205–245.

Chevalier, J. and Mayzlin, D. 2006. "The Effect of Word-of-Mouth on Sales: Online Book Reviews." *Journal of Marketing Research* 43: 345–354.

Dupuy-Salle, M. 2014. "Les cinéphiles-blogueurs amateurs face aux stratégies de captation professionnelles: entre dépendance et indépendance." *Réseaux* 32 (183): 123–159.

Gillespie, T. 2010. "The Politics of 'Platforms'." *New Media and Society* 12 (3): 347–367.

Jurafsky, D., Chahuneau, V., Routledge, B., and Smith, N. 2014. "Narrative Framing of Consumer Sentiment in Online Restaurant Reviews." *First Monday* 19 (4). doi: http://dx.doi.org/10.5210/fm.v19i4.4944.

Kaufer, S. 2011. "Interview with Stephen Kaufer, Tripadvisor CEO." *Preferente.com*, October 22. Retrieved from www.preferente.com/english/marketing-english/interview-with-stephen-kaufer-tripadvisor-cofounder-and-ceo-102405.html (accessed November 20, 2015).

King, R. A., Racherla, P., and Bush, V. 2014. "What We Know and Don't Know about Online Word-of-Mouth: A Review and Synthesis of the Literature." *Journal of Interactive Marketing* 28 (3): 167–183.

Liu, Y. 2006. "Word of Mouth for Movies: Its Dynamics and Impact on Box Office Revenue." *Journal of Marketing* 70: 74–89.

Luca, M. 2011. "Reviews, Reputation, and Revenue: The Case of yelp.com." Harvard Business School Working Paper 12–016.

MacKenzie, D. 2011. "Evaluation Cultures? On Invoking 'Culture' in the Analysis of Behavior in Financial Markets." Working paper. Retrieved from www.sps.ed.ac.uk/__data/assets/pdf_file/0007/64564/EvalCults11.pdf (accessed November 20, 2015).

MacKenzie, D. 2014. "The Formula That Killed Wall Street: The Gaussian Copula and Modelling Practices in Investment Banking." *Social Studies of Science* 44 (3): 393–417.

Mellet, K., Beauvisage, T., Beuscart, J.-S., and Trespeuch, M. 2014. "A 'Democratization' of Markets? Online Consumer Reviews in the Restaurant Industry." *Valuation Studies* 2 (1): 5–41.

Morris, J. W. 2015. "Curation by Code: Infomediaries and the Data Mining of Taste." *European Journal of Cultural Studies* 18 (4–5): 446–463.

Orlikowski, W. J. and Scott, S. V. 2014. "What Happens When Evaluation Goes Online? Exploring Apparatuses of Valuation in the Travel Sector." *Organization Science* 25 (3): 868–891.

Oudshoorn, N. and Pinch, T., eds. 2003. *How Users Matter: The Co-construction of Users and Technology*. Cambridge, MA: MIT Press.

Pharabod, A. S. 2015. "Les avis en ligne: une étape des parcours d'achat." digital-humanities.orange.com, September 2015. Retrieved from http://digital-humanities.orange.com/fr/publications/dossiers/articles/596-4._lire_des_avis_en_ligne__une_etape_du_parcours_dachat (accessed November 20, 2015).

Pinch, T. and Kesler, F. 2010. "How Aunt Amy Gets Her Free Lunch: A Study of the Top-Thousand Customer Reviewers at amazon.com." *Report, Creative Commons*. Retrieved from www.freelunch.me/filecabinet (accessed May 26, 2016).

Reagle, J. M. 2015. *Reading the Comments*. Boston, MA: MIT Press.

Scholz, T. 2013. *Digital Labor: The Internet as Playground and Factory*. New York: Routledge.

Striphas, T. 2015. "Algorithmic Culture." *European Journal of Cultural Studies* 18 (4–5): 395–412.

Surowiecki, J. 2005. *The Wisdom of Crowds*. New York: Anchor.

Tapscott, D. and Williams, A. 2006. *Wikinomics: How Mass Collaboration Changes Everything*. New York: Portfolio.

Vásquez, C. 2014. *The Discourse of Online Consumer Reviews*. London: Bloomsbury Academic.

5 Deconstructing the algorithm

Four types of digital information calculations

Dominique Cardon

Algorithmic calculations currently play a central role in organizing digital information, and in making it visible. Faced with the deluge of disordered and disparate data collected on the web, algorithms form the basis of all the tools used to guide the attention of Internet users (Citton 2014). In turn, rankings, social media buttons, counters, recommendations, maps, and clouds of keywords impose their order on the mass of digital information. For many observers, algorithms have replaced various human editors (journalists, librarians, critics, experts, etc.) to prioritize content that deserves to be highlighted and brought to public attention. Algorithms have thus come to serve as the new "gatekeepers" of public digital space (Zittrain 2006). It is therefore common that criticisms of algorithms reproduce, in a new context, the accusations often leveled at mass media in general: that they reflect the economic interests of the owners, distort markets, ignore the margins, are sensational, conformist, vulgar, etc. It is as if the calculation techniques of the web reflect only the interests of those who program them. But this simple manner of critiquing the power of algorithms neglects the strictly technical dimension of these new gatekeepers, as they make transparent the economic forces that extend throughout the new economy of the web. In this chapter,[1] we argue that we cannot view the new computational techniques of the web as merely reflections of the interests of their owners. Extending the philosophical approach of Gilbert Simondon, we want to explore the technical and statistical properties of these computational tools, focusing particularly on the ways in which they require us to think differently about the production of power and hegemony on the web, and the ways it shapes and orients information online.

The various calculation techniques implemented on the web exhibit great differences that are often effaced by the unifying effect of algorithms. Indeed, there exists a huge variety of ordering and classifying procedures: the search rankings of Google, the reputation metrics of social media, techniques of collaborative filtering, the 'programmatic' advertising of 'real-time bidding' (RTB), and the multiple 'machine learning' techniques that are becoming increasingly widespread in the calculations used by 'big data.' We would also like to clarify the different web calculation techniques in order to describe the digital worlds they give rise to, each according to their own individual logic. Designers delegate values and goals to computer objects that make "cognitive artifacts" (Norman 1991) responsible for operating processes and choices, as well as for authorizing and preventing, for classifying and orienting (Winner 1980; Introna and Nissenbaum 2000). Their development has progressively integrated technical solutions that can address a wide range of problems related to statistics, usage, laws, markets, etc. that appear at various stages. In addition, we would like to investigate the connection between the "mode of existence" of the technical object (Simondon 1989) and the regimes of engagement that determine and promote certain modes of action, hierarchies, and forms of representation in the social world (Introna 2011).

Four types of calculation

In order to simplify the issues governing the classification of online information, it is possible to distinguish between four types of algorithmic calculation within the ecosystem of the web (see Figure 5.1). Metaphorically speaking, we identify these four groups with regards to the place they occupy in terms of the world they are each trying to describe. As summarized in the figure below, the calculations can be thought of as located *beside, above, within,* and *below* the mass of online digital data. Audience measurements, in the first place, are located *beside* the web to quantify the clicks of Internet-users and determine the *popularity* of websites. Second, the group of classifications based on PageRank, the classifying algorithm at the heart of Google's search engine, is located *above* the web, as these calculations determine the *authority* of websites based on the hypertext links that connect them. Third, measurements of *reputation*, developed within online social networks, are positioned *inside* the web, as they give Internet-users a metric that evaluates the *popularity* of people and products. Finally, *predictive* measures that personalize information presented to the user employ statistical learning methods, *below* the web, to calculate the navigation pathways of Internet-users and predict their behavior in relation to the behavior of others with a similar profile or history.

These distinctions between the groups of calculations are largely based on different types of digital data: clicks, links, social actions (likes, retweets, etc.), and behavior patterns of Internet-users, respectively. Thus, they each require different statistical conventions and calculation techniques. In analytically isolating these four means of classifying digital information, we wish to reveal the distinct principles that sustain each type of algorithm: popularity, authority, reputation, and prediction. The hypothesis driving this study on the various measurements of digital information is that these four groups, whose nuances are often poorly perceived and understood, use different ordering principles that each justify different ways of classifying digital information. In mapping these algorithmic calculations, we would like to show that the various technical and industrial issues guiding the digital economy could also be understood in terms of a competition over the best way to rank information.

	Beside	Above 	Within	Below 
Examples	Audience measurement, Google Analytics, advertising	PageRank (Google), Digg, Wikipedia	Facebook friends, Twitter retweets, public opinions	Amazon recommendations, tailored advertising
Data	Views	Links	Links	Behavioral Traces
Population	Representative Samples	Weighted (censal) vote	Social networks, declared affinities	Assumed individual behavior
Form of calculation	Voting	Classification and ranking	Benchmarks	*Machine learning*
Principle	*Popularity*	*Authority*	*Reputation*	*Prediction*

Figure 5.1 Four types of digital information calculations.

The popularity of clicks

The first group of digital calculations is constituted by audience measurements that gauge, *beside* the web, the popularity of websites by quantifying the number of clicks by 'unique visitors.' This measurement is the main unit used to account for the popularity of online media and, through simple equivalence, the advertising revenues they will receive. Audience measurements imitate a democratic vote: each Internet-user who clicks has one (and only one) voice, and the sites that dominate the rankings are those able to attract the most attention. As seen in the history of quantifying the public audiences for newspapers, radio and television, such forms of measurement found legitimacy through their close proximity with democratic procedures (Méadel 2010). Indeed, the 'public' and the electorate are often considered interchangeable collective entities. They share the same idea of statistical representation, founded on the counting of single voices, and both seem to constitute the heart of the idea of a nation. They are both organized around an asymmetry between a small center of "transmitters" (the political sphere, the mediasphere) and a silent population of receivers (electors, viewers). At the center, several media divide up the dispersed individual votes collected by a program, educating and unifying people who share the same experience. Thus popular programs unite a "grand public" by giving rise to an "imagined community" that participates in the formation of a collective civic representation (Anderson 1996).

However, with the increasing deregulation of the media sector and the ever larger role played by advertising, audience measurement has served less to construct a 'public' than to merely gauge 'market share.' In the digital world, where the supply of information is abundant and uncontrolled, the audience has lost all connection with the idea of public political representation. On the web, audiences are measured in two different ways (Beauvisage 2013). First, based on a mass-media model, measurement can be user-centric: a probe is installed in the computers of a representative sample of Internet-users by a measurement company (e.g., Médiamétrie/Netratings or ComScore) that records navigation patterns in order to later classify the audience of the most visited websites. Every month, these sites are ranked by popularity, which determines advertising rates (Ouakrat *et al.* 2010). Already imperfect for measuring television audiences, this method can be even more misleading when applied to the web (Jouët, 2004). The other common technique for measuring web-based audiences (often combined with the first) is more site-centric. Owners of websites can learn about their traffic (for example, with Google Analytics) and combine this information with that of agencies specializing in measuring web audiences (e.g., Médiamétrie, Xitu or Weborama in France). But visitation measurements reveal neither who is behind the computer screen, nor whether the page opened was read or not, nor the socio-demographic characteristics of visitors. So, between user-centric and site-centric methods, a distinct polarity has emerged in terms of the transformation of digital calculations. On one side, traditional marketing professionals are interested in classifying their publics, using the variables produced by both marketing and sociology: profession, income, age, lifestyle, and location. They know a lot about individuals, but little about their behavior. On the other, knowledge of the internet-users' behavior is well recorded using profiles, but little is known about the individuals themselves (Turow 2011).

In response to this dichotomy, Internet pioneers have invented other means for measuring information and distributing it to the public. Indeed, applied to knowledge, popularity is no guarantee of quality; it overwhelmingly encourages conformist and mainstream choices. It largely measures the dissemination of products from a small number of cultural producers to a large and passive public. But as the public becomes increasingly 'active,' there is an increasing demand and desire to find higher quality information. Since the web has greatly disrupted the traditional asymmetry between a (small) supply of information that offers very little variation and a (large) demand that is fulfilled without any real choice, web innovators have developed another classification system not based on popularity, but rather on authority.

The authority of links

When Google launched in 1998 it introduced a new statistical method—on a grand scale—for evaluating the quality of information by locating its calculations *above* the web, in order to record the exchange of 'recognition' signals among internet-users. Unprecedented in media history, this solution was highly audacious. Before Google, the first search engines (Lycos, Alta Vista) were lexical: they ranked websites higher if they included the most keywords requested by the user.[2] The founders of Google opposed this inefficient process with a completely different strategy. Their PageRank algorithm does not try to understand what a webpage is about, but rather measures the *social force* of the sites within the networked structure of the web. Indeed, the particular architecture of the Internet is constructed from a fabric of texts citing each other through hyperlinks. The search engine algorithm arranges information by considering that when a site is linked it is simultaneously receiving a token of recognition, which gives it more authority. On this principle, it ranks websites based on a weighted (censal) vote, which is fundamentally meritocratic: the higher ranked sites are those that have the most hypertext links from sites that themselves have received the most links (Cardon 2012).

The communal, participatory culture of these web pioneers thus broke with the imperative of total representation that traditional media imposed on their notion of the public. In other words, the most visible information is not the most seen, but rather that which Internet-users have actively chosen to recognize by frequently linking to it. Silent viewers are forgotten, as the enumeration of links has nothing fundamentally democratic about it. When a website is cited more frequently by others, its own recognition of other sites has more weight in the calculation of authority. Borrowed from the value system of the scientific community—and particularly from scientific journals, which give more significance to the most cited articles—this measure of recognition has proven to be one of the best possible approximations for judging the quality of information. While researchers and journalists filter information based on human judgment before publishing, search engines (as well as Google News) filter information that has already been published based on human judgments coming from the totality of Internet-users publishing on the web. In the digital sphere, this principle takes the name "collective intelligence" or "the wisdom of crowds" (Benkler 2009). It measures information starting from evaluations exchanged, in a self-organized fashion, among the most active Internet-users and websites. In the same way, many other metrics confer authority to those recognized in communities such as Wikipedia or Digg, in the world of free software, and also in the ranking of avatars in online multi-player games. These platforms employ procedures that allow them to identify the quality of documents or people, independently of their social status, by measuring the authority that they have acquired across the network from the judgments of other users. Through successive approximations and revisions, and often using highly refined procedures, these calculations aim to bring together the reasonability, appropriateness, and accuracy of information with a rigorous conception of rationality and knowledge.

One of the distinctive features of measures of authority is that the signals they record, being placed *above* the web, cannot (easily) be influenced by Internet-users. Indeed, one of the goals of Google's PageRank is that users forget about its existence. In other words, the quality of the measurement depends heavily on the fact that the information being classified and ranked does not act in response to the algorithm's existence. Websites must exchange links "naturally" and "authentically." However, this conception of "instrumental objectivity" (Daston and Galison 2012) is constantly undermined by those who strategically aim to obtain greater visibility on the web. For example, the thriving SEO (search engine optimization) market is made up of companies that promise to improve websites' Google rank. Some offer to refine websites' scripts so that the algorithm can better understand it, while many others merely attempt to construct an artificial authority for the website by using the logic of the algorithm.

Measures of authority reflect the meritocratic culture promoted by Internet pioneers. However, they are subject to two major critiques that have subsequently led to the third group of online information calculations. The first critique is that the aggregation of peer judgment produces a powerful exclusionary effect and a centralization of authority. As with any type of network, whatever is located at the center attracts the most attention and receives inordinate visibility. Because they are cited by everyone, the most well-known websites become the most popular and thus receive the most clicks (Hindman 2009). This 'aristocratic' standard of authority thus devolves into a vulgar measure of popularity. Google has become a powerful magnet for web traffic, allowing the Mountain View company to enhance its revenue by putting advertisements in a separate column called 'sponsored links.' While distinct, these two classifications of links (the 'natural' based on the authority of algorithms, and the 'commercial' based on advertising) comprise the front page for search results on the web, a large intersection for online traffic that disseminates only the most central and conventional websites, as well as those that have agreed to pay in order to be seen (Van Couvering 2008). The second critique concerns the censoring effect of a classification system that uses authority-based measures. The only information that is classified comes from those who publish documents containing hypertext links, like website owners and bloggers; the rest is ignored. But, with the ever-increasing popularization of the Internet, new ways of participating are emerging throughout social networks that are more volatile, conversational, and spontaneous—and less socially selective. Internet-users have become active participants on the web through their Facebook and Twitter pages, and social media have attracted a public that is younger, less educated, and more socially and geographically diverse. Google's algorithm acts as if only hypertext links are able to convey recognition, and their aggregation is based on authority. However, it is no longer possible to do this with 'likes' and pages shared on Facebook, for example. The latter are rather more concerned with subjective meanings, identity constructions, conflicting evaluations, and contextual idiosyncrasies, from which calculations can only derive incomplete or approximate patterns. While hypertext links can appear to project a quantifiable meaning onto the entire web, 'likes' can only give a limited view of the social network of a single person. In addition, social media tend to collapse traditional ranking systems by reorganizing preferences and online patterns around a circle of 'friends' or 'followers' that Internet-users themselves have chosen. While measures based on authority gauge the recognition of documents independent of their authors, it is now users' digital identities that are being evaluated as much as the documents themselves (Cardon 2013).

The reputation of 'likes'

While authority-based measures aim to hide their calculations *above* the web so that Internet-users cannot easily alter or interfere with them, reputation-based measures of social networks are located *within* the web, such that users actively evaluate each other, and can see themselves doing it. The paradigm of these new calculation techniques is Facebook's 'like,' the most visible symbol of a much wider and disparate group of indicators measuring the size of personal networks by number of friends, reputation acquired from published articles and links others have subsequently shared or commented on, the number of times an Internet-user is mentioned in an online conversation, and so forth. Reputation-based rankings measure the users' capacity for having their messages relayed by others. Generally speaking, influence is derived from the ratio between the number of people that a user knows and the number of people who know that user (Heinich 2012), and thus gauges the social power of a name, profile, or image. Competing to have one's arguments validated has become a competition to ensure one's own visibility in the digital sphere. The social web of Facebook, Twitter, Pinterest, Instagram, etc. is full of small assessment and ranking methods, or "gloriomètres," to quote Gabrial Tarde (cited in Latour and Lépinay 2008, 33). They create a landscape with many hills and valleys, a topology that indicates reputation, influence, and notoriety to help people navigate online space. In a world where such counters are ubiquitous, nothing prevents users from acting to improve their rankings. According to authority-based classifications visibility is only ever deserved, but in the reputation-based sphere of social media it can easily be fabricated. Here, self-fashioning a reputation, cultivating a community of admirers, and spreading viral messages have all become highly valued skills. Across the web, under the eyes of everybody, these minor ranking techniques have turned all users into evaluators and classifiers (Marwick 2013). Of course, this metric is not objective; rather, it produces a massive amount of signals, which are then used by Internet-users to orient their behavior and improve their own measures of value.

As original as they are, measures of reputation have also been the object of several lively critiques, especially with the rapid diffusion of social media over the last few years. The first is that, by choosing to disperse visibility throughout a wide range of micro-assessments and counters to challenge the centrality of authority-based and algorithmic measurements, these methods are confining users within a bubble. In other words, by choosing their friends, Internet-users are making homogeneous choices, bringing together people whose tastes, interests, and opinions resemble their own. Consequently, metrics based on affinity create 'windows' of visibility that take on the hue of their own social networks, which prevents them from accessing information that may surprise or unsettle them, or contradict their a priori opinions (Pariser 2011). The second critique is that these many small, local measurements can be difficult to aggregate because of their heterogeneity. There are no common conventions for producing a clear representation of the constant bubbling and buzzing of Internet-users exchanging friend requests, likes, and retweets. The meanings that are trapped within the continual micro-assessments of reputation within social media are too varied, too calculated, and, moreover, too contextual to be truly comprehensible. The grand arena of expression found on social networks highlights the many competing signs, desires, and identities that respond to an economy of recognition and repu- tation. While sincere in their expression, when decontextualized and aggregated these signs can often be considered neither true nor authentic. In a space where visibility is strategically produced, a growing discrepancy is created between what individuals say or project and what they actually do. This gap throws digital calculations off balance, making it difficult to understand the massive amounts of online data. It's unclear whether they should attempt to interpret what Internet-users say—which is very difficult—or merely follow the myriad traces in search of an interpretation—something they are doing better and better.

Prediction through traces

From the latter option, the final group of digital calculations has thus emerged, which exists *under* the web, recording the traces left by Internet-users as dis- creetly as possible. This method is characterized by the use of a specific statisti- cal technique called "machine learning," which has radically shifted the way in which calculations have penetrated our society (Domingos 2015). It aims to per- sonalize calculations based on the traces of online activity to encourage Internet- users to act in one way over another, as seen in the recommendation systems employed by Amazon and Netflix. These predictive techniques have been added to most of the algorithms that measure popularity, authority, or reputation, whereby they *learn* by comparing a user's profile to others who have acted or decided in a similar way. Based on probability, the algorithm guesses that a person may do something that they haven't yet, because those with similar online behavior patterns have done so before. The user's possible future is pre- dicted based on the past actions of similar users. It is thus no longer necessary to extract information from the content of documents, from judgments pronounced by experts, from the size of an audience, from community recognition, or from the preferences reflected in a user's social network. Rather, this method con- structs user profiles based on the traces of online behavior to develop predictive techniques that adhere closer to their actions.

To justify the development of these new predictive techniques, promoters of *big data* have attempted to discredit the wisdom and relevance of human judg- ment. Individuals, they claim, constantly make evaluation errors: they lack dis- cernment, systematically make overly optimistic estimates, are unable to anticipate future consequences by focusing too much on the present, are guided by their emotions, are easily influenced by each other, and lack a well-developed sense of probability.[3] Supported by new findings in experimental psychology and economics, the architects of the latest algorithms suggest that only the real behavior of individuals can be trusted, not what they claim to be or do when experimenting on social media platforms. The global regularities observed throughout the huge number of traces allows for estimations of what users would *actually* do. Thus, predictive algorithms do not respond to what people merely *say* they want to do, but rather to what they really *want* to do, without saying it.

A new way to 'calculate' society

As ever-increasing amounts of data are accumulated, the transformations wrought by big data are, above all, characterized by a revolution in the 'epistemology' of calculations. We would like to highlight three major shifts in the way in which our society represents itself through numbers: (1) a shift in the anthropological capacity of calculations, as these measurements have become much easier to gauge; (2) a shift in the representation of social groups, as categories are increasingly unable to represent those individuals who stand out; and (3) a shift in the social production of causality, as statistical correlations no longer proceed from cause to effect, but rather re-create and estimate probable causes from their effects. These transformations are challenging the long statistical tradition that was constructed, together with the state, to map the nation based on stable conventions and descriptive categories of the social world (Didier 2009). This tradition guaranteed on the one hand a certain degree of consistency and solidity through the "law of averages," and on the other hand, a sufficient legibility to create common categories (Desrosières 2014; Boltanski and Thévenot 1983). But since the early 1980s, society has greatly expanded beyond the categories of those institutions attempting to record, measure, and act on it. Indeed, in an underlying way the current crisis of political representation is bound up with the weakening of statistical forms that once gave structure to the social world. The suspicion among individuals towards the way in which politicians, journalists, scientists, and trade-unions represent them has its basis in the refusal to be locked into predefined categories. So it is precisely to uphold the right to uniqueness and singularity that a widespread reinvention of statistical techniques was initiated to conceive of society, without categorizing individuals too strictly. New digital calculations employ Internet-users' online traces to compare them with other users' traces, based on a system of probabilistic inferences that does not have the same need for statistical information to be plugged back into a highly categorical system. This is no longer thought of as the cause of behavior, but rather as a network of likely attributes estimated based on past behavior.[4] If the 'society of calculation' has so thoroughly penetrated into the smallest aspects of our lives, it is because the social no longer has the consistency that once allowed its representation, using broad and superficial categories to describe individuals.

A more flexible 'real'

One of the first signs that the standard social statistical model is weakening can be observed in the shifting position where data classification takes place, which can be said to have moved three times with respect to the information being measured. It was first located *beside* the web, where the clicks of web-users were counted. Next, it moved *above* the web, forgetting about the users to focus on the signs of authority they exchanged. Then it moved *within* the web itself, to social media where visibility is not based on merit, but rather is a function of self-fashioning and identity construction. Finally, they have shifted *below* the web, as the algorithms, unsatisfied with users' excessive speech, record the online traces of their real behaviors. The trajectory of this shift shows how statistics, once external and distant portraits of society, have progressively come to enter into contemporary subjectivities, comparing behavior patterns before surreptitiously calculating what users are going to do without them knowing. Thus, what was once observed from above, through categories that allowed algorithms to group and unite individuals, is now observed from below, through the individual traces that set them apart. And significantly, this new digital assessment is a form of radical behavioralism that calculates society without representing it.

This trajectory reflects the problem of reflexivity resulting from the intensive use of statistics by actors in the social world (Espeland and Sauder 2007). Unlike the natural world observed by science, human society greatly adapts its behavior to information and the statistics given about it. The scientific ideal of instrumental objectivity is essential for stabilizing "facts" (Daston and Galison 2012), conferring on statistics the confidence and assurance necessary to frame public debate. However, this external measurement position is increasingly difficult to maintain, and the main indicators of social statistics have been accused of misrepresentation (Boltanski 2014). The neoliberal policies introduced in the 1980s have also helped to erode the authority of these categories, by assigning new uses to statistical tools; they now serve less to *represent* the real than actively *act* on it. 'Benchmarking' techniques have contributed to the downfall of the metrics embedded in the social worlds they claim to describe, in addition to the development of 'new public management,' new accounting standards within organizations, and new evaluation and rating mechanisms expanding the use of indices, charts, and 'key performance indicators' (KPIs). Statistical objectivity has thus become instrumental; it is no longer the value of the numbers themselves that is important, but the measured values between them. And, to quote Goodhart's famous law: "Once a measurement becomes an objective, it ceases to become a good measurement" (cited in Strathern 1997). Yet another aim has been assigned to indicators: turning individual social actors themselves into tools for calculation by locating them in environments that tell them how to measure, all while giving them a certain amount of autonomy. But because they are poorly interconnected, this group of indicators does not comprise a comprehensive system. Overall, computational expertise has thus come to replace professional authority in organizing the visibility of digital information—though the fact that these measures can be false is no longer considered to be problematic (Bruno and Didier 2013; Desrosières 2014).

It has become increasingly common to take a specific measurement of a certain activity as a wider indicator of the actual phenomenon being measured; for example, the number of complaints from abused women becomes the actual number of abused women, or the high schools with the best test results become the best schools, and so forth. A performance indicator, often unique, thus becomes a tool for interpreting a much broader context. As well, the reflexive nature of the indicators makes the social actors themselves increasingly strategic, and also renders the 'real' more and more manipulable. Within this context, the latest calculations offered by big data are returning to a more solid exterior position in response to instrumental 'benchmarking' measurements. But they are not supposed to be located beside or above the data being measured, as detached observers surveying the social world from their laboratories. Big data has abandoned probability-based surveys, evaluations of information quality, and has concealed calculations within the black box of the machines, so that users cannot influence or alter it. It has thus reanimated the instrumental objectivity of the natural sciences, but this time without the laboratory: the world itself has become directly recordable and calculable. Indeed, the ambition here is to measure ever closer to the 'real,' in an exhaustive, detailed, and discreet manner.

The crisis of categorical representation

The second transformation that has disrupted the way in which society reflects itself in numbers is the crisis of statistical consistency, which orders a system of categories that maintain stable relationships among themselves. The collection of social statistics no longer adheres to or resonates with actual societies: statistics no longer allow for the representation, from the variety of individual behaviors, of a totality with which people can identify. While statistics have never been more pervasive, they are also increasingly and frequently contested. Global statistical indicators—such as the unemployment rate, price indexes, or the gross domestic product—are often seen as manipulable informational constructs that can be used for a variety of political ends. So they play a dwindling role in the figuration of the social. Under the influence of Rational Expectations Theory, the managerial state has instead reoriented its statistical activities towards econometric methods that aim to formulate and evaluate public policies (Angeletti 2011). Also, within national institutes of statistics, the nomenclature of "occupations and socio-professional categories" has since the early 1990s been gradually replaced by more specific or one-dimensional demographic variables, such as diploma or revenue (Pierru and Spire 2008).

Indeed, econometrics has challenged and marginalized the global statistical models of sociologists in the name of science. With their cross-referenced tables and geometric methods for analyzing data, econometricians prefer linear regression techniques that lead to verification when, 'all things being equal,' there is a correlation between two variables. The data analysis techniques that sociologists and statisticians had established in the 1970s, especially factorial analyses of the principal components, sought to project a set of varied attributes on a two-dimensional plane. Whereas social phenomena were once measured by constructing an overview of society via socio-professional categories, calculations based on 'all things being equal' try to isolate, as specifically as possible, two distinct variables to determine whether one acts on the other, independent from all other variables comprising the cumbersome notion of 'society.' Through mathematics, such econometric calculations individualize the data entered into the models, seeking to make them as exact and unambiguous as possible. This approach is wary of categories that are too broad, which risk contaminating the calculations, creating tautological explanations, and allowing political and social assumptions into the equation. The econometric turn within national statistics has thus laid the groundwork for digital algorithms of big data, based on large numbers. Since computational resources now allow it, it is no longer necessary to refine and limit the models to determine the correlations among variables that serve as assumptions: it is feasible now to ask the machine to test all possible correlations among a growing number of variables.

If calculations appear to have become more dominant today, it is because society has grown increasingly complex and difficult to measure. The logic of heightened customization that has given rise to the current techniques for ordering information is a consequence of the expressive individualization that has accompanied the development of digital applications. Previously, in hierarchical societies where access to public space was highly regulated, it was easy to speak on behalf of individuals using the categories that represented them. Governments, spokespersons, and statisticians were able to 'speak' of society through the aggregations of data they had constructed to depict it. Today, this abstract, disembodied notion of society seems more and more artificial and arbitrary, and less able to represent the diversity of individual experiences. The dispersion of subjectivities through public digital spaces has encouraged individuals to represent themselves (Cardon 2010). The individualization of lifestyles and the augmentation of social opportunities have contributed to an increasing volatility of opinions, a diversity of career paths, and a multiplication of areas of interest and consumption. A growing number of behaviors and attitudes are thus less immediately correlated to the large interpretive variables that sociologists and marketers are more used to. Even if the traditional sociological methods for determining behavior and opinions are far from disappearing, they are no longer able to map society with the same specificity.

Recent developments in statistical techniques have sought to overcome these difficulties by renewing and reinventing both the nature of the data and the calculation methods used. A systematic shift has occurred in the selection of data intended for computers: for more stable, sustainable, and structuring variables that place statistical objects into categories, digital algorithms prefer to capture *events* (clicks, purchases, interactions, etc.), which they record on the fly to compare to other events, without having to make broad categorizations. Rather than 'heavy' variables, they seek to measure signals, actions, and performances.

With this dissolution of national statistical frameworks, most sampling techniques that once allowed researchers to measure a phenomenon within a defined population have been rendered obsolete.

The deterioration of traditional sampling techniques encouraged a radical break in statistical methodologies. Armed with the immense power of computers, big data developers expect comprehensive data sets and are happy to capture them "raw" (Gitelman 2013). Some claim that it is better to collect all data without having to first select or filter through them. But of course, abandoning systematic requirements for data selection in digital calculations has had several consequences (Boyd and Crawford 2011). First, these records concern only those who are active, those who have left traces; others—non-connected, more passive and untraceable—are being excluded from the structures of networked data. Also, the lack of categorical infrastructure for keeping statistical records has contributed to an increasing customization and fragmentation of calculations. In most web services that implement massive data-processing techniques, it is about reflecting back to Internet-users themselves the appropriate or corresponding information. Significantly, the only tool to ensure the widespread representation of data is the map. Geolocation, which allows users to zoom in and out of their own positions, will be the last totalizing tool that remains when all the others are gone (Cardon 2014).

The obscuring of explanatory causes

A third transformation has upset the basis of standard statistical models: correlations do not require causes (Anderson 2008). Acknowledging our ignorance of the causes that are responsible for the individual actions, we have given up looking for an a priori explanatory model. In addition, a new relationship with causality has developed in some areas of statistics, giving 'Bayesian' models a posthumous victory over the 'frequentist' models developed in the tradition of Quetelet (McGrayne 2011). The statistical models of the new data scientists come from the exact sciences, in that they inductively search for patterns by making the least possible number of hypotheses. Current computing power allows for all possible correlations to be tested without excluding any on the grounds that the events leading to them may never come to pass. It would be misleading to assume that these methods search only for correlations 'that work' without bothering to explain them. In reality, they produce many models of behavior that only appear a posteriori, and thus as tangled explanations whose variables act differently according to different user profiles. In a unified theory of behavior, algorithms operate as a continuously shifting mosaic of contingent micro-theories that articulate local pseudo-explanations of likely behaviors. These calculations are intended to guide our behavior to the most probable objects: they do not need to be understood, and very often they cannot be. This inverted way of fabricating the social reflects the reversal of causality effected by statistical calculation to address the individualization of our society, as well as the indeterminacy of an increasingly large number of determinants on our actions. The current logic used by researchers and data scientists is indeed striking in how it attempts to reconstruct frameworks of society: upside-down and from below, starting from individual behavior to then infer the conditions that make it statistically probable.

The examination of computational techniques rapidly unfolding in the digital world today allows us to understand the socio-political dimensions of the choices made by researchers using algorithms to represent society. The principles they implement offer different frames of engagement, according to whether their users prefer measurements based on popularity, authority, reputation, or behavioral prediction. Calculations are thus constructing our 'reality,' organizing and directing it. They produce agreements and systems of equivalence that select certain objects over others, and impose a hierarchy of values that has gradually come to determine the cognitive and cultural frameworks of our societies. However, the choice between these different techniques of representation is not up to the sole discretion of the designers; they are too deeply rooted in the transformation of our societies. Indeed, calculations can really only calculate in those societies that have made specific choices to make themselves calculable.

Notes

1 This text reconfigures the pedagogic typology presented in Cardon (2015).
2 For more on this history, see Batelle (2005) and Levy (2011).
3 See, for example, Ayres (2007) and Pentland (2014).
4 On the "erosion of determinism" in statistical calculations, see Hacking (1975).

References

Anderson, B. 1996. *L'imaginaire national: Réflexions sur l'origine et l'essor du nationalisme*. Paris: La Découverte.
Anderson, C. 2008. "The End of Theory: Will the Data Deluge Make the Scientific Method Obsolete?" *Wired Magazine* 16 (07), July 2008.
Angeletti, T. 2011. "Faire la réalité ou s'y faire: La modélisation et les déplacements de la politique économique au tournant des années 70." *Politix* 3 (95): 47–72.
Ayres, I. 2007. *Super Crunchers: Why Thinking-by-Numbers Is the New Way to Be Smart*. New York: Random House.
Batelle, J. 2005. *The Search: How Google and Its Rivals Rewrote the Rules of Business and Transformed Our Culture*. New York: Portfolio.
Beauvisage, T. 2013. "Compter, mesurer et observer les usages du web: Outils et methods." In *Manuel d'analyse du web en sciences humaines et sociales*. Edited by C. Barats, 119–214. Paris: Armand Colin.
Benkler, Y. 2009. *La richesse des réseaux: Marchés et libertés à l'heure du partage social*. Lyon: Presses Universitaires de Lyon.
Boltanski, L. 2014. "Quelle statistique pour quelle critique?" In *Stat-activisme: Comment lutter avec des nombres?* Edited by I. Bruno, E. Didier, and J. Prévieux. 33–50. Paris: Zones.
Boltanski, L. and Thévenot, L. 1983. "Finding One's Way in Social Space: A Study Based on Games." *Social Science Information* 22 (4/5): 631–680.

Boyd, D. and Crawford, K. 2011. "Six Provocations for Big Data." Paper presented at *A Decade in Internet Time: Symposium on the Dynamics of the Internet and Society*, Oxford, September.

Bruno, I. and Didier, E. 2013. *Benchmarking: L'État sous pression statistique*. Paris: Zones.

Cardon D. 2010. *La démocratie Internet. Promesses et limites*, Paris: Seuil.

Cardon, D. 2012. "Dans l'esprit du PageRank. Une enquête sur l'algorithme de Google." *Réseaux* 31 (177): 63–95.

Cardon, D. 2013. "Du lien au *like*: Deux mesures de la réputation sur internet." *Communications* 93: 173–186. doi: 10.3917/commu.093.0173.

Cardon, D. 2014. "Zoomer ou dézoomer? Les enjeux politiques des données ouvertes." In *Digital Studies: Organologie des savoirs et technologies de la connaissance*. Edited by B. Stiegler, 79–94. Paris: FYP Éditions.

Cardon, D. 2015. *À quoi rêvent les algorithmes. Nos vies à l'heure des big data*. Paris: Seuil/République des idées.

Citton, Y. 2014. *Pour une écologie de l'attention*. Paris: Seuil.

Daston, L. and Galison, P. 2012. *Objectivité*. Paris: Les Presses du Réel.

Desrosières, A. 2014. *Prouver et gouverner: Une analyse politique des statistiques publiques*. Paris: La Découverte.

Didier, E. 2009. *En quoi consiste l'Amérique? Les statistiques, le New Deal et la démocratie*. Paris: La Découverte.

Domingos, P. 2015. *The Master Algorithm: How the Quest for the Ultimate Learning Machine Will Remake Our World*. London: Penguin Random House UK.

Espeland, W. and Sauder, M. 2007. "Rankings and Reactivity: How Public Measures Recreate Social Worlds." *American Journal of Sociology* 113 (1): 1–140.

Gitelman, L., ed. 2013. *Raw Data Is An Oxymoron*. Cambridge, MA: MIT Press.

Hacking, I. 1975. *The Emergence of Probability*. Cambridge, UK: University of Cambridge Press.

Heinich, N. 2012. *De la visibilité: Excellence et singularité en régime numérique*. Paris: Gallimard.

Hindman, M. 2009. *The Myth of Digital Democracy*. Princeton, NJ: Princeton University Press.

Introna, L. D. 2011. "The Enframing of Code: Agency, Originality and the Plagiarist." *Theory, Culture and Society* 28 (6): 113–141.

Introna, L. D. and Nissenbaum, H. 2000. "Shaping the Web: Why the Politics of Search Engines Matters." *The Information Society* 16 (3): 169–185.

Jouët, J. 2004. "Les dispositifs de construction de l'internaute par les mesures d'audience." *Le temps des médias*, 3 (2): 135–144.

Latour, B. and Lépinay, V. A. 2008. *L'économie science des intérêts passionnés: Introduction à l'anthropologie économique de Gabriel Tarde*. Paris: La Découverte.

Levy, S. 2011. *In the Plex: How Google Thinks, Works and Shapes Our Lives*. New York: Simon & Schuster.

McGrayne, S. B. 2011. *The Theory That Would Not Die: How Bayes' Rule Cracked the Enigma Code, Heated Down Russian Submarines and Emerged Triumphant from Two Centuries of Controversy*. New Haven, CT: Yale University Press.

Marwick, A. E. 2013. *Status Update: Celebrity, Publicity and Branding in the Social Media Age*. New Haven, CT: Yale University Press.

Méadel, C. 2010. *Quantifier le public: Histoire des mesures d'audience de la radio et de la television*. Paris: Économica.

Norman, D. 1991. "Cognitive Artifacts." In *Designing Interaction: Psychology at the Human-Computer Interface*. Edited by J. Caroll, 17–38. Cambridge, UK: Cambridge University Press.

Ouakrat, A., Beuscart, J.-S., and Mellet, K. 2010. "Les régies publicitaires de la presse en ligne." *Réseaux*, 160–161: 134–161.

Pariser, E. 2011. *The Filter Bubble: What the Internet Is Hiding from You*. New York: Penguin.

Pentland, A. 2014. *Social Physics: How Good Ideas Spread: The Lessons from a New Science*. New York: Penguin.

Pierru, E. and Spire, A. 2008. "Le crépuscule des catégories socioprofessionnelles." *Revue Française de Science Politique* 58 (3): 457–481.

Simondon, G. 1989. *Du mode d'existence des objets techniques*. Paris: Aubier.

Strathern, M. 1997. "'Improving Ratings': Audit in the British University System." *European Review* 5 (3): 305–321.

Turow, J. 2011. *The Daily You: How the New Advertising Industry Is Defining Your Identity and Your Worth*. New Haven, CT: Yale University Press.

Van Couvering, E. 2008. "The History of the Internet Search Engine: Navigational Media and the Traffic Commodity." In *Web Search: Multidisciplinary Perspectives*. Edited by A. Spink and M. Zimmer, 177–206. Berlin: Springer.

Winner, L. 1980. "Do Artifacts Have Politics?" *Daedalus* 109 (1): 121–136.

Zittrain, J. 2006. "A History of Online Gatekeeping." *Harvard Journal of Law and Technology* 19 (2): 253–298.

6   Baffled by an algorithm

Mediation and the auditory relations of 'immersive audio'

Joseph Klett

You are sitting in a room. On your head is what appears to be a conventional pair of stereo headphones. A brand, *Mantle*, is embossed on the ear cups. The headset is connected to a black box console which in turn connects to a user interface on your computer. From this screen you select "calibrate user." The system generates a series of tones, and you are prompted to locate each tone in relation to your avatar on screen: one tone seems to come from just behind your right shoulder; another appears squarely in front of you, perhaps slightly to the left. You plot a number of tones—maybe 12 in total—then wait a moment while the system creates your profile. The interface indicates the system is ready for you to listen.

Still seated, you select from your library the song "Deja Vida la Volar (REMASTERED)" by Víctor Jara. A media player pops-up with familiar buttons labeled PLAY, PAUSE, ADVANCE, and REVERSE. Something new, you also see a drop-down menu labeled LOCATION; for now you leave this set to "studio." You press play. The strum of Jara's acoustic guitar dances side to side in front of you as the wooden percussion thuds low in the space to your right. Your eyes trace the sound of the Andean flute as it languishes just beyond the reach of your left hand. When Jara's voice appears it looms large, directly in front of you—yet it sounds strangely empty, as if coming from a cavernous place. What you hear resembles what you may have heard in any decent pair of headphones playing digital stereo sound.

Then you move your head. Inside the headset strap, a gyroscope registers your movement. Inside the console, a microprocessor translates this information and reads it using an algorithm that instantaneously transforms the out-going signal to track the movement of your head. You are now looking to your left; the voice that was in front of you can now be heard over your right shoulder. The flute is now just beyond the reach of your *right* hand. The percussion is behind you. You slowly pan your head left to right, and so does the sound, matching your direction to its virtual space. You move your head faster, and the algorithm reroutes through a compression sequence, allowing the signal to transform without audible clipping. Thanks to the perceptual 'cone of the confusion' that sticks directly out of human ears, you don't register the fewer samples in each microsecond of playback.

You venture back to the interface. You find the drop-down menu. You switch LOCATION from "studio" to "CBGB." The sound is automatically transformed from the box studio in which Jara recorded, and into the oblong acoustics of the famous rock venue. It doesn't matter that the New York venue opened two months *after* Jara's tragic death in Chile; another Jara fan online has made the effort of remastering a lossless MP3 of the recording into the object-based format you are hearing now. This format reinterprets the recording as discrete sonic objects that are rendered according to the pre-recorded acoustic parameters of actual spaces. Coupled with the algorithm, you may now use a pair of headphones to hear as if you were present in a number of other places.

This is a fictitious account. But it is also the story told by the engineers who author the algorithms of *immersive audio*.

"What do algorithms do?" Solon Barocas, Sophie Hood and Malte Ziewitz (Barocas *et al.* 2013) offer this question to challenge the functional unity and one-dimensionality of algorithms. Algorithms are not immaterial formulae, but practical expressions that affect the phenomenal world of people. Thus, these phenomenal alterations require scrutiny for their relevance as algorithmic (Gillespie 2014). Rather than assume algorithms to be self-evident mechanisms, we may locate these technical procedures 'in the wild' to understand how they serve to mediate experience. Algorithms are literally and figuratively 'coded' by actors who translate local meanings and values into technical procedures. If only for their ability to adapt should we be curious about the forms of expression and recognition that algorithms mediate. We find such an example in the digital engineering of immersive audio.

In this chapter I take you to a research and development (R&D) lab where audio engineers experiment with psychoacoustic principles and object-oriented computing to reproduce sound in a virtual listening environment. A microprocessor (and its embedded code) mediates between live or recorded sources and a pair of headphones for playback. This sound is transformed as a relationship involving the perceptual features of the listener, the orientation of the listener in space, and the acoustic behavior of an environment independent of that which is occupied. Immersive audio is initially promoted for film, music, and video game applications—though the creation of this new audio format (Sterne 2012) does not disqualify it from any number of 'augmented reality' applications, such as video conferencing and virtual tours. Indeed, today's developers argue that the future of augmented reality depends not on which information can be relayed by this technology, but how the technology will allow new sensory access to all kinds of information. In this spirit, rather than discriminate by application, digital engineers code immersive audio to produce a particular state of auditory relations. These relations emerge from the mediations of personalization, disorientation, and translocation.

To pursue this argument I use ethnographic material to show how mediations emerge as technical decisions and evaluations as engineers work to bring immersive audio to life.[1] Drawing on theory from sound studies and cultural sociology, I argue that the algorithmic transformations of immersive audio provide an effect I call *baffling*: with a double meaning as both material and symbolic confusion, baffling insulates the listener from common acoustic space while rearranging their perception of what is meaningful in that space. This technological act of mediation (in the agentive and encultured sense developed in Appadurai 2015) engages the user's attention while physically blocking other sounds from intervention. To answer the question posed by Barocas and co-authors, what the algorithms of immersive audio do is execute material and symbolic processes which in effect reorganize the auditory relations between listeners and their *im*mediate social worlds.

Mediation

Media technology is today a pervasive arbiter of culture. Audio, for example, provides a means and a medium through which we perceive symbolic material and action. Inside a pair of headphones, we construct a new sonic relationship to the world which overlays meaning (Beer 2007; Bijsterveld 2010; Bull 2012; Hosokawa 2012). The mechanical behavior of that technology thus matters because it colors the symbolic content that it mediates. To this effect, Jonathan Sterne and Tara Rodgers (2011) address the semiotics of algorithmic signal processing, where metaphors of the "raw" and "cooked" obtain to the timbre of audio playback; engineers will use these categories to identify "un-touched" sounds from processed sounds, and thus, in the vacuum of the medium, such symbolic categories allow engineers to specify absolute or "pure" sounds that stand to be processed or left alone.

Beyond audible signals, audio engineers code algorithms to the acoustic properties of sound as well. Sterne (2015) describes the meaning of "dry" and "wet" audio to recording engineers as they combine pre-ordained *signals* and algorithmic *affect*. In this symbolic work, engineers separate the situational effect of reverberation as holding a non-dependent relationship with the absolute object of sound—imagining, in effect, that a sound and that the space of that sound could be effectively divided, extracted from their acoustic situation, and recombined elsewhere. By this logic, a sneeze 'itself' is the same, isolated sound in the vaulted halls of a modern library, in the dunes of a beach, or in the very short distance between headphones and ear canals. From the audio engineering point of view, all that changes is how these spaces alter the perception of that sneeze. For an engineer, this can be an attractive position to take: if sneezes are ontological objects separate from their realization in a perceived space, then an independent association between the two would allow algorithms to manipulate the perception of sonic space during playback. Run any sound through a filter and it is as though you're hearing it coming from a totally different space. Such algorithms thus reconstruct audible 'space' by referencing a set of predetermined acoustic features within a single audio production. In this regard, as Sterne affirms, algorithms produce a representation of sonic space that is built from, at best, partial information about the spaces it claims to represent. For Sterne, "Artificial reverb at once represents space and constructs it" (2015, 113).

Here I should affirm that the symbolic categories which engineers negotiate do not exist apart from the very work of negotiating actual signals. This pragmatic temperament helps us avoid the urge to treat symbols as if they were on a higher plane than their physical expression. In this sense, media historian Lisa Gitelman (2004) cautions against "dematerialization" in theories of media effects: only if we think of media in a vacuum and never in contact with use can we imagine these sounds to be truly "virtual," as though experience is as flexible and interchangeable as text on a blank page. Such reification of symbols-as-text prevents us from understanding what makes one act of mediation more seductive than another. Audio technology is particularly convincing when static signals are replaced with multi-stable systems that provide a context in addition to signals (Langsdorf 2006). In practice, contexts are situations made from a blend of material and immaterial interactions. How we understand a situation—if we recognize it at all—depends on our perception of those interactions. As Martina Löw explains, "a selection has to be made from among the profusion of the perceptible, and that perception is therefore not direct in nature. It merely conveys the impression of directness while being, in fact, a highly selective and constructive process" (Löw 2008, 41). What we regard as the transparency of new media technologies is thus a product of the sensual "atmospheres" (Löw 2008) or "envelopes" (Rawes 2008, 74) that are produced in the contact between encultured listeners and auditory situations.

As material culture, sound is not a passive medium connecting and disconnecting two subjects, for example, in dialog. Rather, sound is an active event in a "hyper-relational world" that constitutes the auditory relations and non-relations of those subjects (Revill 2015). For example, when using the poorly-insulated earbuds that often come bundled with our digital devices, it is others who hear the excess sonic energy leaking from our ear canals and into a common sonic environment. To illustrate this relational aspect of sound, Heike Weber (2010, 346) traces the public history of headphone listening from stationary to portable use, noting the discourse of "respectful listening" that surrounded monaural listening performed with a single ear-bud: respectful because it was discrete to the individual's ear, but also because it left the other ear available to the non-mediated world. Portable, stereophonic headphones would then effectively "privatize" this experience by bringing it within the intimate space of the ear (Weber describes a "cocoon"), while publicly demonstrating the secret sensory access that the headphone user enjoys (Hosokawa 2012). As these studies show, sound allows us to be part of social situations in ways we don't even realize. The more time we spend in this state of using audio technology, the less we are available to those other situations, and the other audiences who co-construct those situations in auditory relations with us.

To capture the material and symbolic dimensions of mediated listening, we must imagine listening to audio not as a binary (on/off) practice, but as a continuous phenomenal relationship with sound in space. To this effect, Tia DeNora (2000) has argued that audio provides a "scaffolding" onto which the self can be latched and upheld, and onto which the actor can be actively entrained to rhythms and affectively swayed by tones; respondents report on the use of audio in everyday life as both a pragmatic technics and a source of self-recognition. Gordon Waitt, Ella Ryan, and Carol Farbotko (2014) call this the "visceral politics" of sound: the practical experience of discourse through sound (in their example, at a parade to raise awareness of climate change) provides a resonance of meaning with sonic material. And while most audio listening involves a piece of music or a soundtrack that the listener selects, Mack Hagood (2011) identifies that listeners are at the same time electing to suppress the ambient sounds outside of their audio devices. In the case of noise-cancellation technology, this ambient sound includes the invert 'noise' produced as an abject artifact around which listeners retreat further into their personal soundscapes. What these pragmatic studies tell us is that perception is a two-way street: the focusing of auditory attention necessarily entails an ignorance of other sensible information in the environment.

Social life is a library of performances organized by genre, and perception is critical for recognizing those genres. In this sense, auditory relations pertain directly to what Ari Adut (2012) terms a "general sensory access" to a common set of experiences. Our phenomenal experience of qualitative information depends on perception as intimately linked to cognition, rather than subordinate to it (McDonnell 2010; Martin 2011; Klett 2014). This means that in direct and unmediated encounters, I may develop new habits of perceiving that attempt to harmonize with my cognitive practices. The more a habit becomes routine, the more likely it will dictate my perception. But when I listen to personal audio technology I am individuated perceptually, not socially. When using headphones with active noise cancellation, a red light shines for others to recognize our inaccessibility despite our bodily presence. Yet to others, I am still accountable for my presence and behavior—my cough, some rustling, a giggle. In this sense of drawing my attention (but not me) from immediate situations, mediation processes do not negate but merely revise my priorities as an actor. It then bears asking: from whence these priorities? If, as sound students tell us, the mediation of auditory relations is a product of cultural codes applied as algorithms, who is writing these codes?

At Mantle R&D

Mantle is an international audio firm with offices in several countries around the globe. Mantle R&D is the firm's Silicon Valley-based laboratory. The lab is composed of a director, eight salaried engineers, and a handful of interns who stay for a season at a time. Their specialties include electrical and computer engineering, sound design, and, of course, writing and coding algorithms.

Most of the engineering work done in the lab is done through a blend of digital and electro-acoustic experimentation. While the image of an R&D laboratory may conjure white coats and cabinets, it is more accurate to picture the 'open office' environment of any software developer or tech startup: long, non-partitioned tables, grey upholstery, standard office doors and windows, posters and various toys strewn about the space. The whole space takes up about 1,200 square feet.

Mantle R&D was created specifically for developing digital signal processing (DSP) for different formats of speaker technology. DSP can be used to do many different operations, affecting any number of dimensions in the production of electronic sound, while wholly constituting the substance of that sound. Thus, while charged with the development of DSP technology in viable applications, this is as much as they are told to do by higher-ups. The placement of the office in Silicon Valley can be seen as an organizational attempt at 'immersion' by dropping a collection of research engineers into a hotbed of digital technology. During my stay in the lab—alternating full-day visits from May to August 2011—the engineers began experimental research in the use of DSP for immersive audio.

Baffling, in three steps

To understand what the algorithms of immersive audio do, I study the process of engineering prior to the final coding of the processor. This labor entails a set of recursive projects conducted over several months. Leading the project is senior engineer Adam and his assistant Stefan, who is working as an intern on his summer break from engineering school.

As Adam and Stefan work to erect an experimental array, gather data, engage scientific literature, and write code, they navigate their way toward the goal of immersive audio through a raft of technicalities that provide practical resistance to reaching their goals. They face regular decision-points in their research where technical know-how can only provide so much; rather, symbolic codes are assigned and assessed in the course of experimentation—decisions that have a direct effect on the function of the technology.

Engineers of digital audio products shape the aesthetic qualities of sound such as color, temperature, and texture. Further, engineers codify particular algorithmic processes that shape the auditory relations between listeners, situations, and acoustic environments. These relations reflect a local set of meanings which the engineers feel are best represented in a set of symbolic transformations of the listening experience. To preview these three mediations:

- *Personalization* defines the listener by a set of auditory features. This allows audio to be customized to a wide range of listener bodies. At the same time the body becomes a part of the audio system, that body is assigned an independent association to the means of reproduction, i.e., the speakers of the headphones.
- *Disorientation* affects a freedom of movement between the orientation of the body within the signal of the audio device. An arbitrary relationship between listener and speaker removes the focal point of conventional auditory fields while lending the listener greater degrees of freedom within the field.
- *Translocation* transforms sound according to the preset features of different acoustic profiles. When the listener is translocated, the acoustic conditions of occupied space need not apply except where they interfere with the signal of immersive audio.

These three concurrent mediations create the desired experience of immersion. Imagined through dialog with the emergence of augmented and virtual reality technology, immersion aspires to transcend the acoustical limits of the phenomenal listening situation. However, this auditory effect depends on a secondary, social function that engineers do not necessarily anticipate. I call this process baffling. Baffling provides the listener a dynamic relationship to mediated sound by disassociating perception from the social situations of immediate listening.

Personalization

Research on perception suggests that variations between user bodies will alter the reception of sound. After all, hearing is a process of distinguishing between changes in audible energy (Evens 2005), and bodies' sense energy in particular places. For audio engineers, these variables include: the size and shape of your ears, the placement of your ears on your head, and the profile of your shoulders. The measured variations between listeners are often referred to as 'individual requirements' for listening. The engineers of Mantle R&D are looking for ways to model individual requirements as algorithmic behavior. This is the process of *personalization*.

At the R&D lab, the engineers' interest in personalization followed a meeting with a professor who studies the psychoacoustic effects of individual requirements. While novel as an engineering application, the professor's work is based on a rather old theory of perception called the head-related transfer function (HRTF). In short, the HRTF is an algebraic representation of the way sound is heard by one ear, then the other, in such a manner that it helps the brain triangulate the source of that sound. By modeling the HRTFs from a sample of different individuals, the engineers at Mantle R&D believe they can write code to process a template that may be modified to fit each user. To gather HRTF data, the engineers use an experimental device called AURA.

> AURA is constructed from 12 speakers suspended on a scaffolding of two horizontal rings. Six speakers sit on stands about four feet high, and six are suspended from the ceiling about ten feet high. Each speaker points toward a chair in the middle. A listener sits in the chair and wears microphones in their ears which point outward to capture sound as it arrives at the ear canal, just past the pinnae. To take a measurement, the speakers play a single tone to produce an even sonic impression. These recordings are captured to Stefan's computer and rendered in visual graphs. Using these graphs he cleans the data to create an audiological map of sound as it reaches the listener.

Most audio engineers would agree that listeners are embodied individuals with unique perceptual apparatuses. But not all agree that every listener should be treated as equal. Prior to immersive audio, engineers had no reason to consider individual requirements beyond those of the engineers who produced the technology. Stereophonic sound, the dominant standard in audio technology, is constituted by two complementary channels of sound which as a whole are evaluated by objective standards of goodness. For example, systems would be evaluated for a high signal-to-noise ratio. A lack of excessive noise (among a list of other variables) would qualify a system as objectively better than others. In relation to these devices of objective quality, professional and self-declared audiophiles reasoned that listeners too could vary in quality depending on how they heard these devices. By contrast, Stefan and the other engineers at Mantle R&D treat individual requirements as integral to the listening experience, and thus, inalienable from a worthwhile digital product.

> After AURA is first assembled in the middle of the lab, Stefan makes some initial recordings to see what the data looks like. The open scaffolding of AURA means that sounds from outside of the device are inadvertently recorded at the same time as the controlled tones produced by the device. The lab is a shared space, where other engineers are conducting unrelated research which nevertheless contributes to the soundscape of the lab.
> Sitting at his computer, Stefan struggles to identify the deliberate tones of AURA from the noisy artifacts that also appear in the data. Without a stable reference point established, he has no clear delineation between what the listener perceives to be noise-as-interference versus noise-as-ambient sound.

Embracing a new epistemology of listening brings with it challenges—especially when listening is conceived as essentially subjective in nature. In the process of personalization, there is no pre-determined or ideal form of playback, as the sound is always, in the end, the subjective experience of the user. Thus, to categorize any sonic artifact as "unwanted" sound would be to arbitrarily over-reduce parts of the signal. Instead, the signal is said to contain its own "good noise" as an audible context in which the listener is situated (Klett 2014). Immersive audio provides both figure and ground inherent to the experience of hearing in space—although this inherently subjective listening experience requires the engineers to make proximate decisions on where another subject's experience begins and ends.

Disorientation

Audio works by transducing electricity into audible vibrations in a diaphragm, or speaker. Because of this, the speaker, as point of contact between system and user, has mostly been imagined as the static frame in which audio sound effectively terminates. Engineers of the stereo era codified the speaker as the objective expression of audio, and the listener as the subject who seeks out this objective expression. Yet the orientation of the subjective listener to an objective sound source inhibits a truly immersive experience of hearing wherein sound terminates in the ear of the listener. For this reason, engineers at Mantle R&D recode the speaker as an intermediary, rather than objective point of orientation. This is the process of *disorientation*.

The art historian Caroline Jones (2005) has referred to "hi-fidelity" stereo systems as the peak of modernist sound: stereo is static, two-dimensional, and ideally in front of the listener—what stereo salespersons would call the system's "sweet-spot." Stereo is a "channel-based" format in the sense that stereophonic signals are captured, stored, and replayed through a single circuit that terminates at the speaker. This assumes a stable, direct, and unmodified reproduction of sound data. From recording to playback, right channel goes to right channel and left channel goes to left channel. The discursive framing of stereo sound as hard-wired to a set of speakers has meant that audio is primarily evaluated using modernist values of objectivity from a stable and unmovable orientation that is always outside of the system being evaluated. Because the system is evaluated as if in a vacuum, the channel-based relationship is ultimately a relationship of sequence, not space.

As one R&D engineer explains, the wisdom of the modern stereo format has been hard to challenge: the biological association of stereo with binaural hearing—that is, sound as perceived by two ears—has naturalized the idea that two-channels are the optimal mode of reproduction. But according to psycho-acoustics, the HRTF describes a relationship between sounds in time *and* space. For example, in cacophonous situations we can isolate the sound of one voice 'beneath' or 'between' many others, or what is aptly called the cocktail party effect. This cognitive process suggests there is more to spatial relationships than just sequence.

> In addition to the controlled tones issued by AURA, other sounds from the laboratory space are recorded inside the open structure of the AURA scaffolding. Indicating the threshold in the data where ambient sounds are interfering, Adam recommends that Stefan aggressively 'truncate' the data so that it shaves microseconds off the recorded data. Truncation offers a more precise representation of how the space is recorded inside of the AURA array. However, precision comes at the cost of accuracy. The digital manipulation of sonic data diverges from the situated action of human perception that it is attempting to emulate.
>
> Adam writes a program to transform recorded sonic data into an algebraic description that may be replicated by an algorithm. Because he is trying to describe a relationship between subject and environment, some known variable must be used in the experiment. Adam instructs Stefan to record only tones of known frequencies. These frequencies will be mapped in visualizations of recorded data, with the expectation that the form these frequencies take can be extracted to leave behind just the auditory relationship of listener to environment that they desire to represent. If they can describe this relationship, then they may manipulate listener perception to any number of orientations in local space.

The data gathered using AURA do not provide stable models, but dynamic spatial relationships which always presume processing to be reproduced. In this sense, AURA treats the listener as an information-processor, and sonic information (conceived in spatial terms) as 'reflections': as signals move from source to perceiver, excess sonic energy keeps moving and 'reflects' off of surfaces which send them back to the listener. This gives sound a sense of sequence and direction—dimensions of a signal that can be manipulated to reorient our sense of where a sound is in relation to another.

To capture the 'good noise' of ambient auditory material, the engineers design AURA to gather data about the listener in relationship to 'the room' as an acoustic space. By reconstructing how we perceive orientation in space, engineers construct a new measure of fidelity. Audio fidelity once referred to verisimilitude derived from a recording or a performance. Verisimilitude assumes a stable, objective standpoint from which performances can be evaluated. By shifting focus from interpretation to orientation, immersive audio refers to the perceptual reproduction of a resonant space with all its acoustic features. Sound is foremost a spatio-temporal issue with no single starting point. In addition to cognitive scientists, it is now computer programmers who are asking, "can one hear the shape of a room?" (Dokmanić *et al.* 2011).

> Stefan spends several eight-hour days adjusting and calibrating the speakers of AURA. Since he is not yet gathering data, and the process is laborious, he uses a rubberized 'dummy' head and torso that has been fitted with microphones in its ears. This allows him to take a series of measures with more care and precision (since dummies don't fidget much).
>
> Stefan records a controlled test tone—it sounds a shrill *beeeeee-yiuuuu!*—playing it through AURA's speakers aimed at the dummy. He repeats this recording throughout the day, at times when the lab is less busy and sounds are steadier. These repetitious sounds—what the engineers call 'good noise'—reveal themselves more readily in the data, and therefore become helpful for marking acoustic territory.

Now with access to cheaper and faster microprocessors for DSP, head-tracking gyroscopes, and global-positioning systems (GPS), digital engineers from outside the stereo tradition have begun to use algorithms to draw the boundary not at the speaker, but at the listener just outside the speaker. If the relation between the recorded sounds and the listener can be calibrated continuously and in real-time, the system would produce an 'independent association' between the sounds and the sources (i.e., the speakers) from which they vibrate. In this relationship, the subject could be free to move with a 360-degree orientation to sound on a horizontal plane. Just as the inclusion of the listener's body provides a heightened sense of being immersed in one's own sonic world, the mediated emancipation of the listener from the static conditions of local situations means that the ear is no longer oriented toward an objectified sound source in the speaker. It is under these conditions of possibility that the algorithm may affect a sense of being in another sonic environment altogether—an ideal of virtualism which until now has evaded audio engineering.

Translocation

Influenced by research on psychoacoustics and the decline of modernism as a design ideology, audio engineers are now exploring a mode of sonic reproduction that is fundamentally user-centric. The way the ears hear, much like the eyes see, is three-dimensional: we perceive sounds as coming from left and right, as well as above and below, near and far. To this end, algorithms approximate these dimensions to affect a sense of reverberation 'within' the spaces in which sonic energy is actually perceived by the ear. At Mantle R&D, this grafting of acoustic features from one space into another is imagined to allow the user of immersive audio to hear as if they are located in a wholly other listening environment. This is the process of *translocation*.

To achieve the sensation of hearing within a different physical environment, the algorithm would arrange a signal according to a model of acoustic responses to an actual space—what engineers call 'the room,' even if it is not enclosed—as built from reference recordings of a controlled tone in a particular environment. As Sterne explains,

> we can think of the sound waves as going out and exploring the space, reaching its outer walls and returning to the center of the space, tracing the territory. That response is then grafted onto the dry signal as if it were in the space.
>
> (Sterne 2015, 125)

In the realm of audio engineering, this act of "tracing the territory" is done by telltale sonic *reflections*. Reflections capture the complex relations between sounds, environments, and the user. In theory, reflections are a byproduct of signals. But as engineers have learned, reflections need not—and very likely *can*not—be avoided when listening to live and technologically-mediated signals. Engineers who say they know acoustics say the physics of a reflection is rather mysterious, but it is not unknowable.

Inside AURA, the controlled signal molds itself around the body in the middle of the experimental array. But sound moves quickly. As the energy strikes the body inside, it bounces beyond the open speaker mountings and strikes the hard surfaces of the room. This sound returns to the experimental space and is recorded as reflections of the original signal.

Visual representations of sound recorded within AURA can be exploded to show the point at which the signal strikes but before any reflections do. But when Stefan takes his measurements, the images he consults show the data is riddled with anomalies. He solicits the help of Dave, who has more experience with acoustical engineering.

Dave suggests taking a series of reference measurements to locate the source or sources of what are probably very early reflections. He enlists my help and, indicating several locations on the apparatus which might be sending this extra information, he and I hold a piece of upholstered paneling while Stefan plays the test tone. We take several recordings, each in a different location, to correlate these physical locations to their location in the visual data.

If the physics of reflection can be stated in an equation, then an algorithm can mimic it. To track reflections in the interest of mimesis, the engineers create a model of the acoustic space that surrounds the body, as opposed to the abbreviated space between an ear and a headphone speaker. As an approximation of an acoustic environment, the experimental conditions of AURA are measured not to the zero-sum ideal of silence, but through a visual representation of recorded reflections in graphic data. By studying the distribution of sound waves charted over time and translating this back to the actual acoustic space within the AURA array, Dave and Stefan divine which sounds are desirable and which are unwanted. Though ultimately imprecise, AURA helps to approximate desirable reflections as relational information, while cutting out the latter as outside this virtual acoustic territory.

Unlike stereo, which 'frames' sound like a portrait within its speakers, immersive audio utilizes reflections to construct a spatial relationship through the arrangement of sounds. As if capturing the territory in a globe, the engineers use digital parameters to overlay any number of territories mapped in a similar fashion, giving the impression of listening to the particular sonic relations found in a totally different environment, with no predefined orientation, and a unique freedom to move. This system is a confluence of mediations—personalization, disorientation, and translocation—none of which takes priority in the sequence of listening. Rather, these processes are always essentially underway in the algorithms of immersive audio.

Algorithms and auditory relations

Social studies of algorithms have tended to focus on code as structured rules that assemble the world as text—numbers, letters, symbols. Less attention is given to how these rules produce relationships in a material sense. But code is only realized in phenomenal media, as associations between the objects they represent and the subjects who perceive them (Coleman 2010). Studies of engagement with digital technologies demonstrate how the technical operation of a device requires a subject in cognitive, sensory, and practical relation (Schüll 2012). In this sense, media technology does not hold unlimited interactive potential. Rather, media, like immersive audio, are co-productive of particular relationships in their use.

For the engineers of Mantle R&D, the body and the space that produces sonic reflections are important because with them, sounds are sonic in the very first instance. In concert with the object-oriented programming of today's virtual and augmented realities, a culture of audio engineering is developing around relational models of perception, inspired by psychoacoustics, cognitive science, and academic work in audiology. This culture is primarily algorithmic, as it is only through deft, high-power computations that our perception may be so effectively mediated. When sounds are physical entities with defined boundaries, engineers may better model perception to an algorithm which reflexively does the work of hearing for us.[2]

> Having calibrated AURA, Stefan is now gathering initial listener data using the dummy torso as his subject. He is busily switching between the computer monitor at his desk and the dummy subject in the center of the experiment.
>   Arriving to gather others for lunch, Mike sidles up and asks, "So how's the guy?" He points to the dummy. In the middle of typing, Stefan looks up with his eyes to the torso slouching slightly in a desk chair. "A bit unstable, but he's okay if nobody touches him."

Sound students encourage theorists to understand audio not as an alternative reality, but as a cultural practice in everyday life. There are few taboos against the use of audio in social spaces. Even group activities with family and friends seem to permit the use of headphones and earbuds. This is the state of auditory relations in today's algorithmic culture. As Brandon LaBelle (2006) argues, auditory relations are irreducible and immutable—but they are transposable. The phenomenal nature of sound constitutes our very sense of occupying space in time, and so by manipulating these elements, we may transform social relations between those habituated to particular technologies, and those who are differently habituated.

Algorithms, like any technology, are inscriptions of social relations. Rather than read the media of algorithms from a distance, I look to their reading, writing and revision in the work of audio engineering. This very sensory practice reveals how the mechanistic procedures of immersive audio begin in the contingent yet value-laden work of a small group of individuals. These audio engineers use algorithms not only to represent the dynamic effect of specific acoustic conditions, but further, to reproduce the relational effect of occupying space as an individual body with unique features of its own. Through the mediations of personalization, disorientation, and translocation, immersive audio directs listener perception away from local situations. Rather than affect a generic auditory space for all listeners to hear, immersive audio processes *subjectivity in independent relation to a non-local space*. The 'object' of immersive audio is then essentially subjective and yet always partial to the perspectives of the engineers.

To say nothing of the socioeconomic inequalities surrounding access to audio technology, I argue that baffling challenges our understanding of 'we-ness' in social situations. Sensory experience is as meaningful as it is shareable (Adut 2012), and audio acts as a "cultural filter" (Blesser and Salter 2006) that blocks certain experiences while encouraging others. The algorithms of immersive audio represent personal differences rather than a shared experience of common phenomena. As these cultural filters become more robust, the immediate experience of local interactions loses power to define the situation. Baffled listeners thus go through a double privatization: the objective isolation of pressing the speaker diaphragm to the ear, and the subjective isolation of interpreting sound made meaningful-to-measure.

In the mediations of immersive audio, the subject's ability to commonly define and distinguish sonic qualities—let alone evaluate these qualities—is reconstructed by personalization, and instead works back on the listener as an object to redefine the situation of listening. This is not to say that people are growing confused or misled by the sensations of immersive audio. Rather, the technology encourages people to opt-out of immediate experience, thus moving to greater levels of individual difference. This creates new challenges for political recognition in complex and stratified societies, where people are cognitively "tuning out" from the situations they occupy (Beer 2007, 858).

Given the uneven distribution of sound pollution across societies and growing markets for digital audio products, silence is quickly becoming a luxury good in the urban environment (Keizer 2010; Hagood 2011; Stewart and Bronzaft 2011; Biguenet 2015). But in this case, 'silence' does not mean quiet 'out there' as much as it represents control over one's hearing 'in here.' For example, the HERE Active Listening is an in-ear device which offers real-time acoustic processing when connected to a smartphone. The device features an equalizer and a set of preset filters with which the user may "curate" their own experience of live music (www.hereplus.me/). As a kind of control by classification, curation suggests individuals may take their pick from a definite range of qualities. This is interesting for the cultural study of algorithms, but not for the *taking their pick* part. Rather, it is the *definite range of qualities* that promise the greatest theoretical traction.

For questions of situational awareness, recognition, and responsibility, algorithms challenge our notions of technology as epistemology. In the practical use of immersive audio, we face an excess of objects at the expense of fewer subjects. Such a state of auditory relations may affect our subsequent states of being in the world. Without speculating on the affective state of empathy among generations of baffled listeners, we can appreciate the fundamental reorganization of auditory relations that occurs in the processes of immersive audio. Even if we are not experiencing a shortage of empathy in society, there remains the more practical issue of our loss of care. And care is a test we can put to our algorithms.

In *Reclaiming Conversation*, Sherry Turkle (2015) urges a new level of interpersonal responsibility in the making of technology. In particular, she suggests that the interactive design of digital technology may be greatly restructured through the conscious inclusion of values in design. This would no doubt provide serious resistance to the baffling that results from immersive audio's latent mediations. But this is exactly the challenge of understanding what algorithms 'do': how can we understand not only the explicit technical function, but the implicit social functions expressed in digital code? As Jonathan Franzen (2015) writes in his review of Turkle's book, "But what of the great mass of people too anxious or lonely to resist the lure of tech, too poor or overworked to escape the vicious circles?" Immersive audio requires little more social solidarity than the user's willingness to calibrate and attach the headset.

There are manifold auditory relations which go unaccounted for in engineering discourse. At no decision point did the engineers of Mantle R&D express interest in baffling the individual; rather, they desire a particular experience of auditory transcendence. The effect of baffling on auditory relations is simply a nonfactor—though it is nevertheless reproductive of a particular relationship between listener and social situation. Can the study of algorithmic cultures identify and operationalize social concepts for engineering new technology? Absolutely. But to the extent that algorithmic cultures are on-going, adaptive processes, we cannot assume that technologies will not continue to baffle, or that users will not desire to be baffled.

Notes

1 The data in this chapter come from a larger research project on culture as it shapes perception in the work of audio engineering (Klett forthcoming). This research received financial support from the National Science Foundation (SES-1128288).
2 It is worth noting that what is 'subjective' here is very specifically defined. The bodies and spaces modeled using immersive audio are static, passive. The individual requirements of a subject are essentially having a torso with shoulders of some discernible breadth, and a head with fleshy ears. Other differences in perception are rendered at some other time, or discarded as unnecessary to the experience of audio. To borrow from Sterne's (2003) history of early audio, the experiences of sound found in so-called 'immersive' audio are often over-reduced to indexicality. In these technologies we find "our ears resonating in sympathy with machines to hear for us."

References

Adut, A. 2012. "A Theory of the Public Sphere." *Sociological Theory* 30 (4): 238–262.
Appadurai, A. 2015. "Mediants, Materiality, Normativity." *Public Culture* 27 (76): 221–237.
Barocas, S., Hood, S., and Ziewitz, M. 2013. "Governing Algorithms: A Provocation Piece." Social Science Research Network. Retrieved from http://papers.ssrn.com/sol3/papers.cfm?abstract_id=2245322 (accessed December 12, 2015).
Beer, D. 2007. "Tune Out: Music, Soundscapes and the Urban Mise-en-scene." *Information, Communication & Society* 10 (6): 846–866.
Biguenet, J. 2015. *Silence*. New York: Bloomsbury.
Bijsterveld, K. 2010. "Acoustic Cocooning: How the Car Became a Place to Unwind." *The Senses and Society* 5 (2): 189–211.
Blesser, B. and Salter, L. R. 2006. *Spaces Speak, Are You Listening?: Experiencing Aural Architecture*. Cambridge, MA: MIT Press.
Bull, M. 2012. "iPod Culture: The Toxic Pleasures of Audiotopia." In *The Oxford Handbook of Sound Studies*. Edited by T. Pinch and J. Bijsterveld, 526–543. New York: Oxford University Press.
Coleman, E. G. 2010. "Ethnographic Approaches to Digital Media." *Annual Review of Anthropology* 39: 487–505.

126   *J. Klett*

DeNora, T. 2000. *Music in Everyday Life*. Cambridge, UK: Cambridge University Press.

Dokmanić, I., Yue, M. L., and Vetterli, M. 2011. "Can One Hear the Shape of a Room: The 2-D Polygonal Case." Paper presented at 2011 *IEEE International Conference on Acoustics, Speech and Signal Processing* (ICASSP).

Evens, A. 2005. *Sound Ideas: Music, Machines, and Experience*. Minneapolis, MN: University of Minnesota Press.

Franzen, J. 2015. "Sherry Turkle's 'Reclaiming Conversation'." *New York Times*, September 28. Retrieved from www.nytimes.com/2015/10/04/books/review/jonathan-franzen-reviews-sherry-turkle-reclaiming-conversation.html (accessed November 1, 2015).

Gillespie, T. 2014. "The Relevance of Algorithms." In *Media Technologies: Essays on Communication, Materiality, and Society*. Edited by Tarleton Gillespie, Pablo J. Boczkowski, and Kirsten A. Foot, 167–194. Cambridge, MA and London: MIT Press.

Gitelman, L. 2004. "Media, Materiality, and the Measure of the Digital; Or, The Case of Sheet Music and the Problem of Piano Rolls." *Memory Bytes: History, Technology, and Digital Culture*. Edited by Tarleton Gillespie, Pablo J. Boczkowski and Kirsten A. Foot, 199–217. Cambridge, MA and London: MIT Press.

Hagood, M. 2011. "Quiet Comfort: Noise, Otherness, and the Mobile Production of Personal Space." *American Quarterly* 63 (3): 573–589.

Hosokawa, S. 2012. "The Walkman Effect." In *The Sound Studies Reader*. Edited by Jonathan Sterne, 104–116. Abingdon, UK and New York: Routledge.

Jones, C. 2005. *Eyesight Alone: Clement Greenberg's Modernism and the Bureaucratization of the Senses*. Chicago, IL: University of Chicago Press.

Keizer, G. 2010. *The Unwanted Sound of Everything We Want: A Book about Noise*. New York: PublicAffairs.

Klett, J. 2014. "Sound on Sound: Situating Interaction in Sonic Object Settings." *Sociological Theory* 32 (2): 147–161. doi: 10.1177/0735275114536896.

Klett, J. forthcoming. "Organizing Sound: An Ethnographic Investigation into the Making of Listening Subjects and Sounding Objects." Unpublished manuscript.

Labelle, B. 2006. *Background Noise: Perspectives on Sound Art*. New York: Continuum.

Langsdorf, L. 2006. "The Primacy of Listening: Towards a Metaphysics of Communicative Interaction." In *Postphenomenology: A Critical Companion to Ihde*. Edited by E. Selinger, 37–47. Albany, NY: SUNY Press.

Löw, M. 2008. "The Constitution of Space: The Structuration of Spaces through the Simultaneity of Effect and Perception." *European Journal of Social Theory* 11 (1): 25–49.

McDonnell, T. E. 2010. "Cultural Objects as Objects: Materiality, Urban Space, and the Interpretation of AIDS Campaigns in Accra, Ghana." *American Journal of Sociology* 115 (6): 1800–1852.

Martin, J. L. 2011. *The Explanation of Social Action*. Oxford: Oxford University Press.

Rawes, P. 2008. "Sonic Envelopes." *The Senses and Society* 3 (1): 61–78.

Revill, G. 2015. "How Is Space Made in Sound? Spatial Mediation, Critical Phenomenology and the Political Agency of Sound." *Progress in Human Geography*. Published online before print February 22, 2015, doi: 10.1177/0309132515572271.

Schüll, N. D. 2012. *Addiction by Design: Machine Gambling in Las Vegas*. Princeton, NJ: Princeton University Press.

Sterne, J. 2003. *The Audible Past: Cultural Origins of Sound Reproduction*. Durham, NC: Duke University Press.

Sterne, J. 2012. *MP3: The Meaning of a Format*. Durham, NC: Duke University Press.

Sterne, J. 2015. "Space within Space: Artificial Reverb and the Detachable Echo." *Grey Room* (60): 110–131.

Sterne, J. and Rodgers, T. 2011. "The Poetics of Signal Processing." *Differences* 22 (2–3): 31–53.

Stewart, J. and Bronzaft, A. L. 2011. *Why Noise Matters: A Worldwide Perspective on the Problems, Policies and Solutions*. Abingdon, UK: Earthscan.

Turkle, S. 2015. *Reclaiming Conversation: The Power of Talk in a Digital Age*. New York: Penguin Press.

Waitt, G., Ryan, E., and Farbotko, C. 2014. "A Visceral Politics of Sound." *Antipode* 46 (1): 283–300.

Weber, H. 2010. "Head Cocoons: A Sensori-Social History of Earphone Use in West Germany, 1950–2010." *The Senses and Society* 5 (3): 339–363.

7 Algorhythmic ecosystems

Neoliberal couplings and their pathogenesis 1960–present

Shintaro Miyazaki

The following chapter argues that theorizing algorithmic cultures as being based upon functional, pure and clean networks, hardware and data systems, might conceal its more dirty, dark, dysfunctional and pathological aspects. These negative aspects are too often downplayed, but certainly belong to the cultures in which algorithms are operatively taking part. An inquiry into algorithms that focuses on these *cultural* aspects should include not only man-made realms, but also the important perspectives of non-human agency, which, while indeed human-made, has gained an effectiveness that far exceeds human culture, communication and aesthetics; self-organized technological processes are an example. Extending the notion of culture towards ecosystem thinking also leads the inquiry to the pathological side of algorithmic culture. Furthermore, an ecosystem is usually considered more living, complicated, sensitive to changes and stimuli than a culture. Its many agents operate in between control and non-control. Algorithmic ecosystems, then, consist of humans and non-humans, respectively non-machines and machines, constantly exchanging signals not only with one other, but also with their environments and their objects, processes, materials and bodies. Such an open-ended approach resonates strongly with MacKenzie's concept of *agencement* as "combinations of human beings, material objects, technical systems, texts, algorithms and so on" (2009, 4).

Along these lines, the first of overall four sections will explain the term *algorhythm*, a neologism that not only enables a specific and at the same time far-reaching understanding of algorithmic culture, namely its rhythmicity, materiality and physicality, but also enables a time-critical approach into its pathologies. The second section will offer short historical accounts of algorithmically-induced resource mismanagement, network breakdowns and other manifestations of unintended bad programming and design, beginning with the 1960s and its first attempts to schedule algorithms in shared systems. It will furthermore explore early issues of the ARPAnet and briefly examine the genealogy of computer viruses. The third section will describe more in detail the so-called AT&T Crash of January 1990, when the long-distance American telephone network was defective for almost a full day and caused considerable economic damage. The final section will develop the notion of neoliberal pathogenesis as a symptom of our current neoliberal, free market society, which, from a media theoretical and historical perspective, is deeply rooted in the above-mentioned context of early distributed networks and computing.

Materiality, algorithms and feedback

Algorithms "bear a crucial, if problematic, relationship to material reality" (Goffey 2008, 16). Rhythm is an important aspect of this relation and a term close to the study of social and cultural phenomena, as sociologists such as Henri Lefebvre have declared (2004). Rhythm is the order of movement, timing of matter, bodies and signals. As it is argued here, rhythms can also illuminate pathological behavior. By interweaving the concepts of algorithm and rhythm, I coined the neologism algo*rhythm* (Miyazaki 2012), in order to shed new light on the more pathological and malicious aspects of algorhythmic ecosystems.

Algorithms are mathematical, symbolic and abstract structures, but should not be mistaken for algebraic formulae. Instructions operated by algorithms are non-reversible, whereas algebraic formulae are reversible. Algorithms are vector-dependent, they need unfolding, and thus they embody time. This crucial difference, as trivial it seems, is of great relevance for understanding our current dependence on algorithmically-driven systems and the ecosystems they affect. In the computational sciences this understanding spread as early as the 1960s with the dawn of higher level programming languages such as Algol 58 and Algol 60. Heinz Rutishauser (1918–1970) was one of the first to use this new form of thinking, which manifested itself in a specific form of notation.

The equal sign is substituted by the so called results-in sign $\Rightarrow$ by K. Zuse, which demands, that the left values shall be calculated as declared into a new value and shall be named as declared on the right side (in contrast to $a+b=c$, which in sense of the algorithmic notation would merely be a statement). (Rutishauser 1956, 28).

Donald E. Knuth (*1938), himself a pioneer in algorithmic thinking of a slightly younger generation and a software historian, also noted that the systematic use of such operations constituted a distinct shift between "computer-science thinking and mathematical thinking" (Knuth and Pardo 1980, 206).

With this in mind, the neologism of algorhythm written with a 'y' (like 'rhythm') might seem redundant. However its potential is not restricted to exposing the time-based character of computation, but also computation's connection to signal processing and thus to rhythms of "spacetimematterings," a term I borrow from feminist theorist and quantum physicist Karen Barad (2014, 168). Machines not only speak or watch one other, but rather—to formulate it in a more technologically accurate way—listen for and detect one another's signals and their rhythms (Miyazaki 2015).

The problematic relationship of algorithms to reality is mediated by signals of mostly electromagnetic waves going through wires, air or another medium. Problems become manifest mostly at passages of transduction from the mathematical and abstract realm to one of the many physical media such as acoustics, optics or electromagnetics. The speed of algorithmic processes, such as algo-trading is not only dependent on the length of the data cables, but since late 2010, on wireless connections established via microwave towers between the sources of the real-time data of financial markets such as the New York Stock Exchange or the Chicago Mercantile Exchange. This tendency proves that algorithms need physics and are still dependent on energy transmissions, even when they are operating on the micro or nano scale.

Nevertheless, algorhythmic activity is most of the time invisible, seamless and unnoticed, and only becomes apparent when the operation causes unintended, unforeseen, pathologic effects. Small programming mistakes, minimal incompatibilities or any sort of carelessly scheduled and designed timing can cause breakdowns. It is highly difficult to program algorithms that can avoid such failures and breakdowns. Great caution is required, especially when the algorhythmic processes are recursively intertwined or create a feedback circuit. Following the work by French epistemologist Georges Canguilhem (1904–1995), pathology is here defined as a divergence from the normal, from a state of equilibrium, that the system faces due to external influences from the environment (Canguilhem 1991, 269). Both living organisms and "technical-economical groups" (Canguilhem 1991, 284)—which I call algorhythmic ecosystems—can show "micromonstrosities," as a result of "false rhythm" (Canguilhem 1991, 276). Due to the physics of telecommunications these micro-failures are not immediately tangible and remain often unnoticed, until they generate concrete, mostly economic, effects. The following sections will offer short historical inquiries into some crucial aspects of situations where capitalistic values are coupled with media technological environments in the period between the 1960s and the late 1990s.

Distributed dysfunctionality

With the dawn of algorithmic thinking, higher programming languages, and operation systems in the early 1960s, the first experiments in data networking, memory sharing and distributed networking and computing soon began. From the very beginning of data communication over computer networks, so-called deadlocks or lockups appeared in the context of memory and storage allocation. These were some of the first occurrences of an algorhythmic ecosystem trapped in feedback loops. Scheduling mistakes of storage sharing were caused for example by conflicting tasks each waiting for the other to release some memory space (Coffman *et al.* 1971, 70). Such simple mistakes could be avoided by preemptive memory allocation and better designed scheduling, but still could not have been anticipated before they actually came to pass. These initially harmless computer failures gained momentum, when bigger networks were affected.

Activated in the late 1960s, ARPAnet was the first attempt to build a massive computer network, which soon reached geopolitical scales. It was the first "large-scale demonstration of the feasibility of packet-switching" (Abbate 1999, 7). As in all following systems of distributed networking, a message in the

ARPAnet was divided into smaller packets that were then individually transmitted over the network towards their final destination. When the topology—that is, the structure—of the network changes due to a failure or an interruption somewhere on their way, the packets can take alternative and different routes. After arriving at the final destination, they are re-assembled into the right order. In this algorhythmic ecosystem the slow emergence of distributed breakdowns began.

In the 1960s and 1970s, electrical and computer engineers-turned-economists and -systems thinkers such as Jay Forrester (*1918) made system dynamics popular. This new approach and research field was an extension of the field of cybernetics, founded by Norbert Wiener (1894–1964), and aimed at creating abstract models of all kinds of systems such as supply chains networks, industrial systems, urban society, even the entire earth, by dissecting the essential processes into positive and negative feedback loops. These were the first attempts to rationalize and measure all human processes, which resonated with the upcoming ideology of late capitalism and neoliberalism. The valuation of everything starts with measurement. At the same time Thomas Schelling (*1921) was researching the effects of micro-behavior in urban communities, specifically analyzing how small, unintended decisions could cause social effects such as race segregation (Vehlken 2015). Ironically, such effects of complex distributed operativity had already been observed and treated within the field of distributed networks. The rigidly programmed micro-operativity of one-node computers in the ARPAnet—which were called Interface Message Processors—affected the overall macro-behavior of its entire algorhythmic ecosystem. Local failures could have "global consequences" (McQuillan *et al.* 1978, 1805).

Leonard Kleinrock (*1934) was one of the first researchers who had to deal with the complexity of distributed networking. After a decade of research he noted in 1978 that not only was the demand process "bursty," it was also "highly unpredictable." Both the exact timing and the duration of network demand was "unknown ahead of time" (Kleinrock 1978, 1321). Kleinrock recognized that the probabilistic complexities of distributed networks are "extremely difficult" and that an effective "flow control" within the network was an important requirement that had at first been underestimated and ignored (Kleinrock 1978, 1322). Such flow control mechanisms could have prevented many deadlock phenomena such as "reassembly lockup," "store and forward deadlock," "Christmas lockup" and "piggyback lockup" (Kleinrock 1978, 1324). But even the most sophisticated algorithmic implementations could not prevent new mistakes. Detecting errors was a part of research and opened opportunities for optimization.

A program called "Creeper" (Schantz 2006, 74), created in 1971 by Robert H. Thomas of Bolt, Beranek and Newman, a company that provided a substantial part of the technological infrastructure for the early ARPAnet, yields more aspects of distributed dysfunctionality within algorhythmic ecosystems.

> Creeper is a demonstration program, which can migrate from computer to computer within the ARPA network while performing its simple task.

> It demonstrated the possibility of dynamically relocating a running program and its execution environment (e.g., open files, etc.) from one machine to another without interfering with the task being performed. Creeper led to the notion that a process can have an existence independent of a particular machine.
>
> (Sutherland and Thomas 1974, 23)

Creeper was one of the first computer viruses emerging in the disguise of a demonstration and utility program that could replicate and transfer itself within its own ecosystem of distributed networks. Creeper was a sort of algorhythmic organism. Similar experiments were conducted in the late 1970s by John F. Shoch and Jon A. Hupp from Xerox Palo Alto Research Center (Parikka 2007, 241). Shoch and Hupp experimented with "multimachine worms" as they called them (1982, 173). They playfully programmed algorhythmic agents, later dubbed Xerox Parc Worms, that showed messages, loaded pictures or operated as alarm clocks in the affected computers. These were computer programs consisting of several algorithms that distributed themselves over a network of computers, and they could algorhythmically orchestrate programmed operations where all parts worked together. Like their predecessor, the programmers were soon confronted with the unexpected, autonomous and pathological behaviors that their algorithms developed.

> Early in our experiments, we encountered a rather puzzling situation. A small worm was left running one night, just exercising the work control mechanism and using a small number of machines. When we returned the next morning, we found dozens of machines dead, apparently crashed.
>
> (Shoch and Hupp 1982, 175)

In order to prevent such incidents, they included a function to stop the activity of the worm—"an emergency escape" (Shoch and Hupp 1982, 176).

Over the course of the 1980s similarly playful experiments were copied, repeated and re-invented independently by many young programmers. With the introduction of the World Wide Web around 1990 by Tim Berners-Lee (*1955) and the concomitant spreading of program-based organisms, computer viruses, as they were later called, became part of a global and an ever-evolving algorhythmic ecosystem. It was not purely by accident, but rather a symptom of the increasing global connectivity of humans and machines via communication and media technology, that around the same time that Berners-Lee programmed and optimized his first web browser at CERN in Switzerland, one of the first major algorithmically-induced economic and technologic catastrophes happened on the other side of the Atlantic.

AT&T Crash 1990

On January 15, 1990 the long-distance North American telephone network, AT&T, crashed and remained dysfunctional for nine hours. Half of the long-distance calls during this time remained unconnected, which led to a financial loss of more than $60 million excluding consecutive economic damages for AT&T (Peterson 1991, 104). This incident, later called the AT&T Crash, was popularized by Bruce Sterling, a science fiction author and pioneer of digital cultures, in *The Hacker Crackdown*, his first nonfiction book published in 1992 (Sterling 1992). The reason for this crash was a small programming mistake implemented via a software update of the newer and software-based telephone line switching-computers located in different centers spread all over the USA. The update was made in December 1989 on all of the 114 switching-computers.

The whole system worked well until on January 15, 1990. At about 2.30 p.m. a computer located in New York started to detect some malfunctions in its system. For reasons still not completely known, this led to a self-triggered shut down for four to six seconds (Neumann 1990, 11). To maintain the connectivity of the network system, this machinic incident, which usually caused no issues and happened frequently, had to be communicated to the neighboring switching-computers. After the New York switching-computer was online again, it sent out messages to other computers closest to itself. This initiated an update of their routing-maps, so that the machine that was previously shut down could be added to the network again. The crucial point is that the malfunction became effective exactly during this updating process. The update of December 1989 made this process 'vulnerable.' Incoming update messages from the previously shut down New York-based computer disrupted its neighbors' seamless operativity, which caused data damage. Each neighboring computer, in turn updating its routing map, shut itself down and after recovering would try to reconnect to the network, thus causing other shut downs (Neumann 1990, 11–13). The whole network was trapped in an eternal algo*rhythmic*, distributed and polyrhythmic refrain: a nation-wide, transatlantic feedback loop.

Two levels of rhythmicity constituted the specific algo*rhythmic* interplay and machinic over-functioning. On one side, there were the rhythms happening within one switching-computer itself, and on the other side, there was the rhythm and timing of the notification messages, which were sent out from one switching-computer to all the computers in its vicinity. The cause of the malfunction was later recognized as a simple programming mistake in the programming language C. In a long "do-while-loop," there were many switch-instructions and in one of the many branches a switch-clause had an "if-loop" in it. This would not have been a problem had it not been for the additional break-instruction, which in the programming language C does not make a lot of sense (Neumann 1990, 13). The program did not act in the way intended and the rhythms concerning its computational operations were thus slightly out of order and started to operate pathologically. The breakouts of loops or the breaks of switch-instruction chains were timed in unintended rhythms. Its algorhythmics were stuttering in a way. These "evil" and "bad" (Parikka and Sampson 2009, 11) rhythms were usually not detectable by the self-monitoring operating systems, but became effective during the abovementioned updating process of the routing maps.

Neoliberal couplings and their pathogenesis

The spread of neoliberalism beginning in the 1980s, and its connection with parallel developments such as the growth of media networks, the Internet, and distributed computing, as well as the emergence of the "digital economy" and "immaterial labor," has already been discussed and treated many times from different perspectives (Castells 1996; Terranova 2004, 75; Hardt and Negri 2004, 187; Pasquinelli 2015). A deeper inquiry into the circuitry and algorhythmics of these media ecological transformations offers insights into the pathologies of late capitalism or neoliberalism, and epistemological coupling with the above described dysfunctionalities in the early history of distributed networks.

Neoliberal thinking and acting, as defined by sociologist and anthropologist David Harvey (*1935), not only aims at maximizing the reach and speed of transmissions and actions within a market, but also

> seeks to bring all human action into the domain of the market. This requires technologies of information creation and capacities to accumulate, store, transfer, analyze, and use massive databases to guide decisions in the global marketplace. Hence neoliberalism's intense interest in and pursuit of information technologies.
>
> (Harvey 2007, 3)

Not only is the privatization, commodification and financialization of everything a key trait of neoliberalism, but also the "management and manipulation of crises" (Harvey 2007, 160–162). Furthermore, as social activist and writer Naomi Klein (*1970) controversially claims, there are undeniable relations between Milton Friedman's influential economic theories of free market, more precisely the belief in laissez-faire government policy, and many critical events in global politics (Klein 2008). Regarding the dominance of Friedman's economic theory and the Chicago School of Economics, it is therefore not by accident that not only financial markets, but also economies in general, are confronted with disasters and bursting economic bubbles. These economic disruptions are effects of accelerated positive feedback circuits, which emerged due to deregulation. According to neoliberal thinking, it is not desirable to control the numerous feedback loops of a market. Financial liquidity is desired and a seamless flow of data is required. Control mechanisms and opportunities to regulate feedback processes are consequently deliberately left out. While the algorithmic cultures of neoliberal societies have been programmed to operate within enforced conditions of accelerated "competition" (Davies 2014, 39) the likelihood that such complicated and highly coupled ecosystems will collapse is increased. These collapses, crises and breakdowns are effects of the neoliberal coupling of distributed networking with capitalized values, which lead to economic pathogenesis.

Using a model discussed by renowned economist Hyun-Song Shin (*1959), German cultural theorist and historian Joseph Vogl (*1957) compares the self-amplifying feedback circuits of financial markets with the so-called Millennium Bridge Problem, where "tiny, haphazard oscillations" at first caused imperceptible synchronization of pedestrians' steps, which, via positive feedback, led to synchronized lock-step motions and the dangerous swinging of the whole bridge (Shin 2010, 1–6; Vogl 2014, 116). Banks and agents in financial markets act similarly. The movement of the bridge provides a model for how price changes in the market and the succeeding adaptive synchronizations and their feedback effects lead to breakdowns and pathological effects in the market. Following Hyman P. Minsky (1919–1996) Vogl also argues that economical crises and breakdowns are not merely effects of external influences or by political decisions, but are rather results of the behavior and "endogenous" activity of the financial economy itself.

> Unlike cybernetic and self-regulating systems, the financial market is inclined to be spooked by its own tranquillity and destabilized by its own stability. The very efficiency of its functioning turns out to be utterly dysfunctional.
>
> (Vogl 2014, 119)

Breakdowns, accidents, and catastrophes thus belong to the functioning of financial markets. Neoliberalism is thus an important driver and source of the genesis of socioeconomic, network-based pathologies. It is furthermore crucial to distinguish different degrees of dysfunctionality. Pathologies, breakdowns and dysfunctionality within such algorhythmic ecosystems are, as the historical examples have shown, often not dirty or dark in the direct sense, they are usually very ordered, since they are triggered by algorithms and machinic processes. Still, their effects on human society are more than harmless. The scope of these damaging effects can vary from small-scale infrastructural issues concerning only a few companies or universities, to large-scale problems affecting entire nations, and even global communication systems.

In the case of breakdowns and catastrophes in financial markets due to algo-trading such as the Flash Crash of May 6, 2010, the damages were not only financial; they also appeared in the emotional and mental conditions and bodies of the human agents who were coupled with its networks (Borch *et al.* 2015). Market crashes induced by algo-trading show, in impressive and frightening ways, the consequences of deregulated, still highly connected and ultra-fast operating algorhythmic ecosystems becoming accidentally caught in uncontrollable positive feedback loops, leading to financial losses or gains within minutes or seconds. Other more minor crashes or errors belong to the daily business of distributed networking. Notably, each failure serves as the beginning of a further optimization of neoliberal deregulation. The neoliberal system, to this end, is in a "continual state of pathological excitability" (Massumi 2014, 332).

The pathogenesis of neoliberal economies is due to their enforced coupling with distributed networks, which enable the commodification, valorization and financialization of everything. This coupling of monetary values with real world signals and data generates a manifold of real-world social problems, from financial losses due to network breakdowns, losses sustained by distributed agencies in online markets, crashes in financial markets, to broader events such as the financial crisis of 2008, the current (2009–) Greek government-debt crisis and more. It is highly upsetting that such historical events are part and parcel of neoliberal strategies. We are not only living in a "society of control" (Deleuze 1990), but in a society of the controlled planning of un-controlled crises.

Conclusion

When an algorithm is executed, processes of transformation, and of transduction from the mathematical realm into physical reality, are involved. These processes are not trivial. They have been designed to appear simple, but the becoming of an algorithm, its unfolding and metamorphosis into an algorhythm, often involves issues, problems, frictions and breakdowns. As has been described in this chapter, distributed computing and networking makes these frictions much more effective since algorithms need to operate together, interact and share resources with one another via a distributed interplay of agents. As the examples earlier have shown, sometimes these acts of coordination, togetherness and self-organization can spin out of control. In neoliberal contexts, where algorithms—as in algo-trading—are programmed to compete with one other and where all transactions are tightly connected to one other, these frictions, catastrophes and perturbations become day-to-day business.

As described above, the language of engineering in context of emerging network management in the 1960s and 1970s was tightly coupled to economic thinking, using terms such as resource, demand or allocation. The technical coupling of financial flows and algorhythmic ecosystems is from monetary perspectives much more pathologic. It evolved a decade later with the dissemination of computer technologies from the 1980s onwards. The AT&T Crash of 1990 is an example of an early pathological, algorhythmically-induced incident.

The focus on the dirty, dark, dysfunctional and pathological aspects of algorithmic culture reveals an increased need for open, ecologic, non-objectifying, non-human, signal- and rhythm-based approaches, and shows that the developments in the 1960s and 1970s which are outlined above are crucial for more in-depth inquiries into the pathological side of contemporary neoliberal society. Algorhythmic ecosystems evolved in the 1960s with the need to share resources, such as the allocation of data memory space. Interconnectedness grew soon and in the early 1970s engineers were confronted with the problem of harmless local failures building up, unforeseeably, to global large-scale issues, such as network breakdowns. In the late 1970s and early 1980s, software engineers programmed distributed algorithms with counterintuitive, emergent behaviors that they compared to viruses. The AT&T Crash of 1990 stands for the beginning of the digital age, where the vulnerability of a fully algorithm-based and networked society became drastically manifest.

Theorizing global communication networks, and their couplings to economic, social and cultural environments, as processes of algorhythmic ecosystems, exposes the rhythmic activity that underlies these networks, which are constituted by flows of affective, energetic events. These rhythms can cause positive and negative, pathologic effects depending on their operative role in feedback circuits and ecosystems of heterogeneous agencies.

This chapter shows furthermore the risks when abstract worlds such as those of algorithms are coupled with the world of capitalistic values and monetary machinery. These developments are long-term effects of the problematic relations inherent in algorithms from the very beginning. In order to unfold, algorithms need embodiment, for instance, via low-energy signals. And in the age of neoliberalism even those low-energy signals are part of a monetary system. Algorhythmic ecosystems are growing and becoming increasingly interconnected. Thus it is of great importance not only to understand the positive, useful and pleasurable sides of algorithmic cultures, but also to obtain an accurate and comprehensive overview of their pathological sides as well. This chapter is only the starting point for this difficult endeavor.

References

Abbate, J. 1999. *Inventing the Internet*. Cambridge, MA: MIT Press.

Barad, K. 2014. "Diffracting Diffraction: Cutting Together-Apart." *Parallax* 20 (3): 168–187. doi: 10.1080/13534645.2014.927623.

Borch, C., Bondo Hansen, K., and Lange, A. C. 2015. "Markets, Bodies, and Rhythms: A Rhythmanalysis of Financial Markets from Open-outcry Trading to High-frequency Trading." *Environment and Planning D: Society and Space*. Published online before print August 21, 2015. doi: 10.1177/0263775815600444.

Canguilhem, G. 1991. *The Normal and the Pathological*. New York: Zone Books.

Castells, M. 1996. *The Rise of the Network Society: The Information Age: Economy, Society, and Culture*. Malden, MA: Blackwell.

Coffman, E. G., Elphick, M., and Shoshani, A. 1971. "System Deadlocks." *ACM Computing Surveys* 3 (2): 67–78. doi: 10.1145/356586.356588.

Davies, W. 2014. *The Limits of Neoliberalism: Authority, Sovereignty and the Logic of Competition*. London: SAGE.

Deleuze, G. 1990. "Post-Scriptum sur les sociétés de contrôle." In *Pourparlers (1972–1990)*, 240–247. Paris: Les Éditions de Minuit.

Goffey, A. 2008. "Algorithm." In *Software Studies: A Lexicon*. Edited by Matthew Fuller, 15–20. Cambridge, MA: MIT Press.

Hardt, M. and Negri, A. 2004. *Multitude: War and Democracy in the Age of Empire*. New York: The Penguin Press.

Harvey, D. 2007. *A Brief History of Neoliberalism*. Oxford: Oxford University Press.

Klein, N. 2008. *The Shock Doctrine: The Rise of Disaster Capitalism*. New York: Picador.

Kleinrock, L. 1978. "Principles and Lessons in Packet Communications." In *Proceedings of the IEEE. Special Issue on Packet Communications* 66 (11): 1320–1329.

Knuth, D. E. and Pardo, L. T. 1980. "The Early Development of Programming Languages." In *A History of Computing in the Twentieth Century: A Collection of Essays with Introductory Essay and Indexes*. Edited by N. Metropolis, J. Howlett, and Gian-Carlo Rota, 197–273. New York: Academic Press.

Lefebvre, H. 2004. *Rhythmanalysis: Space, Time and Everyday Life*. Translated by Stuart Elden and Gerald Moore. London and New York: Continuum.

MacKenzie, D. 2009. *Material Markets: How Economic Agents Are Constructed*. Oxford: Oxford University Press.

McQuillan, J. M., Falk, G., and Richer, I. 1978. "A Review of the Development and Performance of the ARPANET Routing Algorithm." *IEEE Transactions on Communications* 26 (12): 1802–1811. doi: 10.1109/TCOM.1978.1094040.

Massumi, B. 2014. "The Market in Wonderland: Complexifying the Subject of Interest." In *Timing of Affect: Epistemologies, Aesthetics, Politics*. Edited by Marie-Luise Angerer, Bernd Bösel, and Michael Ott, 321–338. Zürich: Diaphanes Verlag.

Miyazaki, Shintaro. 2012. "Algorhythmics: Understanding Micro-Temporality in Computational Cultures." *Computational Cultures: A Journal of Software Studies* 2.

Miyazaki, S. 2015. "Going Beyond the Visible: New Aesthetic as an Aesthetic of Blindness?" In *Postdigital Aesthetics: Art, Computation and Design*. Edited by David M. Berry and Michael Dieter, 219–231. New York: Palgrave Macmillan.

Neumann, P. G. 1990. "Risks to the Public in Computers and Related Systems." *ACM SIGSOFT Software Engineering Notes* 15 (2): 3–22.

Parikka, J. 2007. *Digital Contagions: A Media Archeology of Computer Viruses*. New York and Bern: Peter Lang.

Parikka, J. and Sampson, T. D. 2009. "On Anomalous Objects and Digital Culture: An Introduction." In *The Spam Book: On Viruses, Porn, and Other Anomalies from the Dark Side of Digital Culture*. Edited by Jussi Parikka and Tony D. Sampson, 1–18. Cresskill, NJ: Hampton Press.

Pasquinelli, M. 2015. "Italian Operaismo and the Information Machine." *Theory, Culture & Society* 32 (3): 49–68. doi: 10.1177/0263276413514117.

Peterson, I. 1991. "Finding Fault." *Science News* 139 (7): 104–106. doi: 10.2307/3975512.

Rutishauser, H. 1956. "Massnahmen Zur Vereinfachung Des Programmierens. (Bericht über Die 5-Jährige Programmierarbeit Mit Der Z4 Gewonnenen Erfahrungen)." In *Elektronische Rechenmaschinen Und Informationsverarbeitung*. Edited by A. Walther and W. Hoffmann, 26–30. Braunschweig: Friedrich Vieweg und Sohn.

Schantz, R. E. 2006. "BBN's Network Computing Software Infrastructure and Distributed Applications (1970–1990)." *IEEE Annals of the History of Computing* 28 (1): 72–88. doi: 10.1109/MAHC.2006.4.

Shin, H. S. 2010. *Risk and Liquidity*. Oxford: Oxford University Press.

Shoch, J. F. and Hupp, J. A. 1982. "The 'Worm' Programs: Early Experience with a Distributed Computation." *Communications of the ACM* 25 (3): 172–180. doi: 10.1145/358453.358455.

Sterling, B. 1992. *The Hacker Crackdown*. New York: Bantam.

Sutherland, W. R. and Thomas, R. H. 1974. "BBN Report 2976 of Dec 1974, Natural Communication with Computers. Final Report – Volume III, Distributed Computation Research at BBN October 1970 to December 1974." BBN. Prepared for Advanced Research Projects Agency, December 1974.

Terranova, T. 2004. *Network Culture: Politics for the Information Age*. London and Ann Arbor, MI: Pluto Press.

Vehlken, S. 2015. "Ghetto Blasts: Media Histories of Neighborhood Technologies Between Segregation, Cooperation, and Craziness." In *Neighborhood Technologies: Media and Mathematics of Dynamic Networks*. Edited by Tobias Harks and Sebastian Vehlken, 37–65. Zürich and Berlin: Diaphanes.

Vogl, J. 2014. *The Specter of Capital*. Stanford, CA: Stanford University Press.

8 Drones

The mobilization of algorithms

Valentin Rauer

For centuries technological devices were seen as no more than passive counter-poles to active human subjects, used by them as tools, instruments, and other means to ends. This binary order is now beginning to crumble (Latour 2005; Schulz-Schaeffer 2007), and there are technological objects in existence that can no longer be thought of merely as passive instruments. The use of the 'drone' metaphor to refer to unmanned vehicles marks this transition: drones stand for a shift in which hybridity is increasingly replacing the binary order that previously divided the world into the opposing realms of subject and object.

Why did the drone metaphor appear when it did—in the early 2000s? I will argue that the shift was triggered by a specific connection that enabled us to translate algorithmic methods and algorithmic inscription into action by material objects. Deployed in a vehicle, algorithms are no longer a means of selecting and generating information or knowledge; they are a way of mobilizing and activating objects. Algorithms are thus the core mechanism by which objects are transformed into interactive agents and the drone metaphor is an attempt to flag up this change to a new public order of things.

With this in mind, when I look at the socio-cultural impact of algorithms here, I do so not, as many current studies do, from the point of view of their use on the World Wide Web (Gillespie 2010; Bennett and Segerberg 2012; Gerlitz and Lury 2014; Sanz and Stancik 2014), but in the context of spatial-material devices and mobilizations (Introna and Wood 2004). Mobilized algorithms have societal effects; they intervene—literally—in public spaces and in the spatial preconditions that shape social interactions and social situations. Taking this as my starting point, I consider mobilized algorithms from three angles.

I begin by looking at the kinds of algorithms that enabled spatial mobilization prior to digitalization and the existence of computer-based networks. Specifically, I consider the so-called radio algorithms used in air- and sea-based communications. In these classic mobile environments, algorithms set up interactive communications networks, designating the roles of 'sender' and 'receiver' and defining situations and responsibilities. The use of algorithms to structure interactive networks thus pre-dates digital and web-based connective action.

I then focus on the mobilization of digital algorithms and, after that, on the increasing spatial autonomy of military drones (Sparrow 2007; Schörnig 2010;

Matthias 2011; Strawser 2013; Chamayou 2015). In relation to the first of these issues, I look at the débâcle involving the German defense ministry's Euro Hawk project in the 2000s, the prime cause of which was the failure to build algorithmic autonomy into the device. This case shows us that mobilized objects will fail in their mission if non-human algorithmic actions are ignored and actor status and subjectivity are ascribed only to the human agents involved in their operation (i.e., screen-based remote controllers). In relation to the second issue, I look at the lawsuit brought before the Administrative Court of Cologne by the relatives of Yemenis killed in a drone attack. The relatives accused the German government of complicity in the deaths on the grounds that it had allowed the U.S. military to use a satellite relay station on German territory (at Ramstein) to carry out the operation. Both these cases show the effect which infrastructures and algorithms have in determining actor boundaries and subjectivity: in the Euro Hawk example they are 'zoomed out,' in the Ramstein example 'zoomed in.'

The aim of this chapter, in short, is to establish a sociological concept of mobilized algorithms that goes beyond mere mathematical or digital software or web applications and construes algorithms as mechanisms for setting up 'inter-objectivity' (Rauer 2012a). In this sense, algorithms provide an infrastructure for objects to interact more or less autonomously. In the imagination (Rauer 2012b), this autonomy might be equated with free will or independent intentionality, but such an identification would be misleading.

Algorithms in mobilized environments

In public discourse the term 'drone' is used to denote physical devices and robotic systems that act and make decisions without direct human intervention. It covers a wide range of implements, from simple remotely operated gadgets with minimal autonomy to robotic devices that are able to interact with their environment independently of human control or supervision. Many of the devices involved are simply versions of the 'remote control' planes, trains, boats, and cars of the past. Why do we now refer to such objects as 'drones' or 'unmanned area vehicles'? Is this just a promotional ploy by industry? Or a bid for attention by the media, ever on the look-out for something new? Sociologically, the metaphor is a significant one, because it signals that the devices in question are mobilized and activated by algorithms that confer on them the status of hybrid actors.

The concept of the algorithm has so far attracted little attention in the classic social sciences. Even where digital navigation is under discussion, it often goes entirely unmentioned (November *et al.* 2010). Where it does figure in sociological studies, it is usually referred to as a formal or functional tool in the context of IT or the history of computing (Heintz 1993). In line with more recent approaches (Beer 2009; Vries 2010; Introna 2011; Marres 2012; Knowles and Burrows 2014), I here argue that algorithms should cease to be thought of merely in terms of rational decision-making and digital machine-based intelligence and should instead be understood as a specific mode of communicative action that structures interactions in spatially extended, widely dispersed situations.

Algorithms are part of a broader array of performativities that includes, for example, rituals, narratives, and symbolic experiences (Alexander 2006a). Thus, as documented by Tarleton Gillespie (2014), algorithms assess public relevance, include or exclude political subjects through indexing, anticipate cycles of attention, promise objectivity and impartiality, shape political practice by adapting it to the relevant algorithms, and present the public with a quantified picture of itself. Algorithms do not just produce, select, and aggregate data; they also structure societies' cultural self-representations and create webs of meaning. Like symbols, myths, and ritual performances, algorithms generate motivations, organize experience, evaluate reality, and create social bonds among actors.

The term 'algorithm,' although closely associated with mathematics and information science, is generally defined simply as "a procedure, a method, or a list of instructions for solving a problem" (Simonson 2011, 93). Key to this definition, therefore, is the fact that it is not based purely on mathematics and does not simply depict algorithms as "paper machines" of the sort developed in computing (Heintz 1993, 63ff.). In this broader sense, the term 'algorithm' encompasses a whole range of interactive methods and procedures for solving specific problems. What method is used, and what list of instructions is followed, is not determined a priori. The method may be digital (Anderson 2013), but it can just as well be analogue. Algorithms are not confined to the realms of information technology or online practice. Nor do they have to be reduced to the bare bones of a programming language. In sum: algorithms have to do not only with specific scientific practices but with problem-solving in general—which may, but does not have to, involve mathematical, digital, or technological processes. Although this point is frequently made in abstract terms (Vöcking 2011; Gillespie 2014, 1), practical examples are thin on the ground—a deficit which I hope will be addressed by the examples that follow here.

Algorithms in maritime radio communications

The appearance of algorithms in situations of object interaction pre-dates the invention of unmanned area vehicles (UAVs). Manned aviation and seafaring are two areas in which they have long been used to solve problems of communication and interaction. In these contexts, it is not possible, as it is in everyday face-to-face interaction, to use deictic gestures to ensure the transmission of a unique speech-act to a particular potential responder. When a speech-act is transmitted by radio, everyone in range of the station hears it, but the identities of the speaker and of the intended addressee are unclear. The prime role which Wittgenstein (1969) attributed to deictic gesture in the creation of certainty is inapplicable in these kinds of interactive situations, where the interactants are not in close proximity to one another and speakers cannot use gesture to indicate to whom their speech-act is addressed. It was to solve this problem of deictic uncertainty that radio algorithms were developed. They perform the functions exercised by deictic gestures in face-to-face situations.

As indicated, among the deictic problems associated with 'blind' interaction at distance is the inability to unambiguously assign the speech-act or designate its addressee. To overcome this, the auditory message must be structured according to an 'algorithm of acquaintance.' This delimits not only who is speaking but also what sort of addressee the speech-act is directed at and what content and degree of urgency are involved, thus ensuring those not addressed do not disrupt the interaction. Certain algorithms are designed to ensure further transaction of the interaction—in other words, its mediation to more distant third-party inter-actants. In these cases, the algorithms are known as 'relay algorithms' (a 'mayday relay' is a case in point) and they communicate to the addressee that the speech-act involved relates not to the actor making it but to a third party who is unable to communicate across the distance in question.

These algorithmic conventions generate the communicative certainty else-where provided by indexical markers. More specifically: a radio emergency call issued at sea may be heard by many parties, but if the call does not follow algo-rithmic rules, nobody may feel compelled to respond. This is what social psy-chology calls 'diffusion of responsibility': the greater the number of (in this case) recipients of a message, the more likely it is that all the recipients will assume the message is meant for someone else and, consequently, that no one will take any action to help (Latane and Darley 1968). Such a situation does not just hamper effective action; it is actually dangerous. Radio algorithms address this problem by specifying a particular communication procedure which, despite non-visibility, uniquely defines speakers, addressees, and responsibilities. Thus, the transmitting party must begin by clearly identifying themselves and then proceed according to the format laid down in a pre-defined communications index.[1] Formulations here might include "Mayday," "Pan Pan," "Sécurité," or "Mayday relay" followed by the name and position of the issuing vessel or air-craft and the nature of the assistance required. Prefacing a message with one of these formulations defines the social situation uniquely. An introductory "Mayday relay," for example, establishes that a potentially life-threatening emer-gency is underway involving a third party from whom the speaker has received a distress call. Because of the formulation, all relevant parties know that the speaker does not themselves have an emergency and is acting as an intermedi-ary. As indicated, one of the tasks fulfilled by the algorithm is that of designat-ing, from among multiple potential speakers and respondents, those that will act. In this way, performative responsibilities—in the literal sense of the duty and right to make a response—are generated within a group of unknowns, producing a coercive pressure that obliges someone to speak and someone to respond. This points to the role of algorithms as intermediaries that define speaking agency and response agency.

Despite the fact that the recipients and dispatchers of the message are anonymous and invisible to one another, an entire, complex communicative infrastructure is created simply through the use of the algorithm in question. This precludes any unwanted debate about meaning: all recipients of the message know at once how they are to respond and what further sequence of actions and algorithmic speech-acts they must initiate. Diffusion of responsibility is thus avoided. One algorithm determines the next and this sequence creates a situation of inter-subjectivity that functions without deictic assistance. Inherent in each of the algorithms is a process that simultaneously precludes spontaneous deliberative reasoning and dictates specific responses.

Although sociologically speaking, the spatially extended, widely dispersed communications scenario created by radio technology represents a standard micro interaction, the indexical gestures that provide certainty in such interactions are here absent and are replaced by algorithms. On the face of it, algorithmically regulated communications such as these recall Herbert Blumer's "symbolic interaction" (Blumer 1969). However, that concept was concerned not so much with solving the problem of diffuse responsibility as with identifying interactively generated conventions and schemes of interpretation that facilitate mutual understanding.

In short, then, the classic form of algorithm allows one to establish reference networks through which extended spatial interaction sites can be created and bridged. Such algorithmically generated networks are not confined to the realm of web-based or digital technology; rather, they are a general type of communicative action that structures and intermediates interaction.

In their exploration of spatially extended structures of interaction, Knorr-Cetina and Bruegger (2002) coined the term "global microstructure," which Knorr-Cetina later followed up (2009) with the more abstract notion of the "synthetic situation." Knorr-Cetina argued that synthetic situations go beyond relational networks because they are structured in ways similar to micro-sociological situations in which interactants are co-present. In Knorr-Cetina's account, this "structural surplus" is supplied by screens which, for example, display real-time market activity or provide sites for direct interaction on-screen or via chatrooms. The name which Knorr-Cetina gives to the latter sites is "scopes" (Knorr-Cetina 2005). Although appropriate and convincing in general, use of this terminology in the case of algorithms would be misleading in that it focuses on the surface elements of interaction sites—namely, screens and screened content—rather than on the intermediaries and infrastructure that enable the social relations to deploy. It looks at the endpoints and results of the transaction chain and disregards the intermediate algorithmic operations. What I identify as mobilizing algorithms would, in this scheme, be excluded from the spectrum of sociological enquiry.

The promise of autonomy

The increasing absence of humans in the algorithmic interaction of objects creates the impression that there are already, or eventually will be, fully autonomous algorithmic agents (Suchman and Weber forthcoming 2016). Concerns about autonomous weapon-systems, for example—viewed by computer scientists as "the Kalashnikovs of tomorrow"[2]—have triggered a number of political campaigns. In another quarter: in 2014 the Defense Advanced Research Agency (DARPA) in the United States invited research proposals for the development of autonomy algorithms that would enable small UAVs to move around inside buildings and under the sea without relying on GPS signals or human remote control. The so-called Adaptable Navigation System (ANS) would, it was reported, incorporate "new algorithms and architectures for … sensors across multiple platforms [and] extremely accurate inertial measurement devices that can operate for long periods without needing external data to determine time and position."[3] Even new devices such as the one described will never be fully autonomous (the UAV in question will operate independently only for restricted periods of time). Nonetheless, in situ they will give the appearance of human-like, intention-driven agents, opening up a cultural space of imagined algorithmic autonomy.

Algorithms that rely on measurement devices thus enhance the capability to engage in autonomous interaction and many are already built into civil aircraft and long-distance U.S. Airforce drones. Trends in algorithmic design still envisage on-screen control—or, more pertinently, remote control—by humans, but the human role is shifting from one of operating and guiding the mobile object to one of merely defining aims and destinations and monitoring and controlling algorithmic completion. The inter-object action relies on algorithmic interpretation of data delivered by measurement devices in situ. With the addition of so-called learning algorithms, the devices in question could operate on a non-linear regression model, in which parameters for in-situation decisions are developed autonomously. 'Learning algorithms' such as this create a problem which Matthias has dubbed the "responsibility gap": they produce outcomes and decisions that are not ascribable to specific human agents (Matthias 2004; also Arkin 2007). This gradual trend towards the removal of humans from the loop (Singer 2015) conjures up imaginings of fully technologically autonomous agents acting in quasi-human ways (Lucas 2013). In the terms of Ingo Schulz-Schaeffer's "ascribed agency" (2007, 2009) and the well-known Thomas Theorem which it echoes: if the autonomy ascribed to objects is seen as real, it will be real in its consequences.

Within learning algorithms—such as the Google search facility—ever greater degrees of autonomy are gradually realized. This kind of algorithm operates by non-linear regression analysis, which computes correlations. These point to possibilities of and probabilities about causal relations but not to causality as such. The situation is the same as with the popular 'theory of the stork': the data may show "a significant correlation" between numbers of storks and numbers of births, but this obviously does not prove that the birds deliver the babies (Höfera *et al.* 2004, 88). Where computed statistical models include big numbers, however, the performative appearance is such that, in public discourse, search outcomes are often misconstrued as conveying something about causation, objectivity, and truth. Gillespie (2014, 2) discusses this problem in terms of the "promise of algorithmic objectivity," meaning the tendency of information-seeking algorithms to convey an impression of impartiality even though their results may be coincidental (on the history of objectivity, see Daston and Galison 2007). Here again, the Thomas Theorem comes to mind: if the outcome of a search algorithm is seen as based on causation, it will operate as such in its consequences. Similar performative effects are produced in the case of algorithmically mobilized objects: the instant reaction of these to movement and interference creates the impression that they are autonomous, intentional actors and thereby renders them such in their consequences.

Given the difficulty of empirical analysis here, my point of access in what follows is the classic one of conflict and contestation (Latour 2006), because it affords a rare opportunity to gain entry into social infrastructure that usually operates below the sociological radar (Junk and Rauer 2015). In line with this approach, my chosen examples are the failed Euro Hawk project and the legal proceedings concerning the Ramstein relay station and its part in deaths allegedly caused by U.S. drones.

The Euro Hawk case

During the 2000s, the German defense ministry was looking to update its reconnaissance aircraft and took the decision to replace the old planes with UAVs. Because Germany had no home-grown drone-development program, it bought a U.S. prototype based on the American 'Global Hawk' UAV. The plan was to use this as a platform onto which European-made reconnaissance technology would be built. When the first prototype of the so-called 'Euro Hawk' was ready, it took off from a base in California bound for Bavaria in Germany. Even this first outing proved a fiasco. Because the necessary flight permits had not been granted, the drone set off not eastwards but westwards, towards the Pacific Ocean. It then turned north, flew parallel to the coast in the direction of Canada, and crossed eastwards over uninhabited Canadian territory, across Greenland and the Atlantic, until it reached German airspace and its final destination of Manching, in Bavaria. The permits had been refused because the aircraft was not equipped with a 'sense and avoid' system for preventing in-air collisions.[4] The automatic algorithmic technologies built into these systems establish interconnections between multiple aircraft moving in one another's vicinity. Bearings and altitudes are then fixed algorithmically between the craft, which climb or drop as necessary. The lack of 'sense and avoid' equipment was not the only potential source of danger on the maiden flight. The journey was also marked by repeated loss of contact between the aircraft and the team remotely controlling it in Germany. At these times, the drone was flying unguided and completely blind, posing a real threat to anything in its vicinity.

Faced with these worrying complications, the defense ministry cancelled the project. Estimates of the losses incurred range from €0.5 to 1.5 billion.[5] In the aftermath, a parliamentary committee was appointed to "investigate the way in which the Federal Government had handled the Euro Hawk drone project from the contractual, legal, financial, military, technological, and political points of view."[6] The débâcle, it seemed, was due to inexperience in algorithmic interaction technology coupled with poor political practice. Obviously, practical contact with the devices in question had been limited up to then and no 'best practice' codes had yet been developed. Among the many witnesses and experts questioned by the committee, a number made reference to 'sense and avoid' technology, explaining the various possibilities in this regard. According to one expert, says the committee in its report: "The current conditional approval only applies to Category 2 traffic. [...] By contrast, approval of an unmanned device for all-air operation requires [the inclusion of] an anti-collision system."[7] This quote highlights the key role played by algorithms in the mobilization of objects: if a drone lacks an autonomous interaction-algorithm, the greater part of the airspace is closed to it because the risk of its causing casualties is too high. What might seem to be just another debate over an air-safety device is in fact a pointer to a societal and cultural transformation (Daase 2016). Mobilizing algorithms have a sociological as well as a technological impact because they alter core societal visions of how ordinary situations of interaction are structured. Traditionally, humans sitting in front of screens were regarded as essential to the proper operation of systems of responsibility and liability; nowadays, algorithms are taking over this human function. Screen-based humans continue to monitor operations and control the means and ends of interactions, but it is the algorithms themselves that do the interacting. Algorithms replace subjectivity with objectivity.

The ultimate cause of the Euro Hawk project's failure was thus a missing algorithm—one that would have enabled the drone to communicate and interact with other AVs/UAVs autonomously. Instead, the device continued to depend entirely on remote control by humans.[8]

As explained earlier, algorithms structure communication and interactivity in situations that preclude the use of deictic tools. They convey content and establish rules for speakers, responders, listeners, and others. Just as deictic aids like gesture are key to *face-to-face* interaction, so algorithmic methods are crucial in what might be termed *face-off-face* situations. The absence of a human actor here opens up new possibilities as regards the ways in which societies ascribe social, political, and cultural agency. What is said to do the acting in these situations is an algorithm, but this does not mean that action is no longer ascribed to humans. What it means, rather, is that what is regarded as an 'agent,' 'agency,' or 'actor' is assuming an increasingly diffuse and hybrid shape. How a diffuse and hybrid agency of this kind was called to account before a court of law is documented in the following section.

The Ramstein case

After 9/11, the United States began to make more and more use of drones in the so-called 'war on terror' (Schörnig 2010; Williams 2013; Strawser 2013; Chamayou 2015). The U.S. administration initiated two distinct drone programs: the first, run by the military, was designed to operate solely in war zones; the second, under the aegis of the Central Intelligence Agency (CIA), was intended for use both within these zones and beyond them (Sterio 2012). From the point of view of international law, this latter program blurs the notion of war. In contrast to the military deployment of drones, the use of these devices by the intelligence agencies is kept secret and the resultant lack of transparency and accountability present problems for democratic public spheres and civil societies. In the words of Blum and Heymann, the CIA program, which allows for so-called 'targeted killing' operations, "[displays] more than any other counter-terrorism tactic … the tension between addressing terrorism as a crime and addressing it as war" (2010, 150).

In recent years, a number of non-governmental organizations (NGOs) have responded to the accountability problem by launching a counter-program of 'strategic litigation' (Fuchs 2013). NGOs contact victims of drone attacks, or their families, and encourage them to bring proceedings against drone-deploying states. The aim is not to secure legal victory but to set a public agenda and initiate debate. The victims' stories and the images of violence which the media publish as a result of the cases are of major significance to the public sphere in democratic civil societies (Alexander 2006c; Giesen 2006). The public narration of alleged violations of human rights enables society to regain a degree of discursive power and events come to be defined as 'wrong'/'immoral' or 'illegal'/'illegitimate' (Mast 2011). The litigation is thus both an instrument for mobilizing the public and a performative strategy aimed at rendering hidden victimhood public. The key aim of litigation strategists is not to win the cases but to get the courts to take them on.

NGOs such as the London-based Reprieve group[9] and the European Center for Constitutional and Human Rights (ECCHR)[10] in Berlin have made a number of attempts to get the courts to assume jurisdiction over cases of targeted killing by drone. In 2015, their efforts finally paid off and a German administrative court in Cologne took on a case brought against the German government by a group of Yemeni victims of U.S. drone strikes. This was a remarkable decision: how was it that a German court was prepared to assume jurisdiction in a case that did not involve German citizens and in which the events did not take place on German territory?

The answer lies in the fact that part of the algorithmic infrastructure of the U.S. drone program is based in Ramstein, Germany. According to the whistle-blower and former U.S. drone pilot Brian Brant, the Ramstein airbase is used by the American military as a satellite relay station for directing the movements of combat drones launched in the United States. It was thanks to this disclosure by Brant that the Cologne court accepted jurisdiction over the Yemeni case. Although the plaintiffs' call for Germany to close down the base was ultimately dismissed by the court, the objective, as previously mentioned, was not to triumph legally but to raise public awareness.

Various extracts from the court proceedings encapsulate the respective positions of the plaintiffs and the defendant.

The plaintiffs: In his statement to the court, the plaintiff Faisal Ahmed Nasser bin Ali Jaber explained the basis for his involvement in the action against the German government:

Waleed Abdullah bin Ali Jaber, my nephew, and Salim Ahmed bin Ali Jaber, my brother-in-law, were killed in a drone attack on our village on August 29 2012. Five days before he was killed, Salim had preached against al-Qaeda in his Friday sermon.

...

Because no one takes responsibility for drone attacks and there are no publicly available centralized records, it is impossible to determine how many drone attacks have taken place. It is also impossible to tell exactly how many people have died and whether they were the intended targets of the attacks. The whole process is shrouded in mystery. That is one of the reasons why drones instill such fear: we never know when or whom they are going to attack, or why anyone has even been classed as a target. As a result, it is impossible to work out how to protect oneself or one's family.... It is as if we were living in a never-ending nightmare that we cannot wake up from.

...

The drones keep on flying and threatening our lives. My family is still afraid of the drones. Since the attack that killed Salim and Waleed, my daughter is frightened of loud noises and has been traumatized by the experience. Like so many others, she lives in constant fear.[11]

This statement clarifies the plaintiff's relationship to those who were killed and conveys the permanent situation of threat that prevails at the local level. It speaks to the issue of the violation of the fundamental rights of Yemenis living under the menace of attack by drones. This kind of declaration is a prerequisite to any German court's assuming jurisdiction over a case that does not involve German citizens. Because fundamental rights are rights that are enjoyed by all alike, the expectation is that where such rights are involved, the threshold for acceptance of jurisdiction will be very low. That said, German courts will not take on every case of alleged violation of fundamental rights: even where the victims of drone strikes have been German—as was Bünyamin E., for example[12]—judges have sometimes ruled the suit to be outside their jurisdiction. What made the difference this time was, as previously indicated, the involvement of Ramstein, and a sizeable passage in the court's ruling addresses the role of the airbase as part of the mobilizing infrastructure of a global drone network. It states:

[The plaintiffs submit that] the drone pilots were located in the U.S.A. The data was transmitted via fiber optic cables from the U.S.A. to Ramstein (Germany) and radioed on to the drones by means of a satellite relay station. The drone pilot in the U.S.A. was in constant contact with colleagues in Ramstein. Because of the curvature of the earth, direct control of the drones from the U.S.A. was not possible without the satellite relay station in Ramstein.[13]

At work in this scenario is a micro-global actor with a reach extending from the American continent to Germany. This actor, however, is not reducible to a single person seated in front of a screen. Rather, it encompasses both the pilot in the U.S. and, via interconnecting cables, the latter's colleagues in Ramstein. Because of the geography, the action depends on the mediation of the satellite relay station. The implication is that if this dispersed actor-network perpetrates a violation of human rights and an outside actor who has the possibility and capability to disable the network's intermediary node remains passive and does not intervene, that actor can be held responsible for failing to prevent an illegal act. The algorithmic device is part of the globally "distributed action" (Hutchins 1995). Its location is of significance because it confers responsibility on the host state. If that state—which in the present example would be Germany—is aware of the actor-network and fails to take action against an offending locally deployed device, it is open to prosecution. Awareness is an important requirement of impeachability and this explains the care which the Yemeni plaintiffs take to point out that: "The satellite relay station was constructed in 2010, a fact about which the U.S.A. had informed the defendant."[14]

The court proceedings also make explicit mention of the use of algorithmic analysis of meta-data in targeted killings. They report the plaintiffs' claim that: "Since 2012, so-called 'signature strikes' have also been carried out, in which targets are selected according to particular patterns of behavior and meta-data, without any concrete knowledge as to which actual person is involved."[15] When asked by a journalist whether individuals were assassinated on the basis of meta-data, former head of the U.S. National Security Agency (NSA) General Michael Hayden replied: "[That] description … is absolutely correct. We kill people based on metadata" (Ferran 2014). Despite securing this mention in court, the issue of signature strikes and meta-data did not play a major role in the ultimate outcome.

The plaintiffs' position, as summed up by the court, was that if the evidence showed that the drone war violated international law,

> it followed that the defendant had a duty not to allow it to be conducted from its territory. In this connection, [the defendant's] contention that it had no definitive knowledge was irrelevant. Finally, responsibility was not excluded by the fact the U.S.A. acts on its own authority and as a sovereign entity: there was sufficient opportunity for the Federal Republic to exert decisive influence.[16]

The plaintiffs do not question the legality of active operations; rather, they charge Germany with passivity. Toleration of an algorithmic infrastructure supportive of illegal acts is claimed to be unlawful. A relay station thus features here as a node of radio communications and an algorithmically constituted form of interaction. Among the plaintiffs' demands are "the withdrawal of the allocation of radio frequencies for radio traffic from the satellite relay station at the Ramstein airbase [and] the termination of the usage agreement on the Ramstein airbase."[17]

The defendant: The defendant's justification for its passivity, as summarized in the court's ruling, was that responsibility for micro-global chains of action lay not with Germany but with the U.S.—in other words, in the place where the actors doing the commanding and controlling were located. What mattered, it said, was not intermediate points, but the location from which the drones took off. "The U.S. government, with whom it maintained an intensive dialogue, had [it said] always asserted that no drones were commanded or controlled from Germany and that Germany was not the point of departure for the drone attacks."[18] The argument, in short, is that because the UAVs do not begin their journey in Germany, the German state has no responsibility. Actors are only accountable for what is materially visible and manifests itself in the movement of material objects from A to B. It is only A and B that matter here; the spatial range and mobilizing infrastructure between them do not count as part of the action.

Another thrust of the defendant's argument was that Germany does not have the power or duty to control interactions between other sovereign states:

> German audit powers [it said] would not allow any monitoring of communications data. Moreover, it was not the defendant's job to act as a global prosecutor vis-à-vis other sovereign states. It was the U.S.A. and Yemen that were the acting parties here and therefore the only states with responsibility.[19]

The defendant sought, with some difficulty, to construct a concept of actors and actions that was to its advantage. By invoking the notion of the sovereign state, it shifted the onus onto a collective actor, excluding algorithmic intermediaries from the picture and thereby obviating any ascription of responsibility to itself.

Finally, the defendant raised the issue of intention and motive as preconditions of responsibility:

> The use of Ramstein [it said] likewise did not impose on the Federal Republic any international legal responsibility that would require the defendant to make greater efforts to clarify facts than it had done previously. According to the International Law Commission's draft code, international legal responsibility for assistive actions required positive knowledge and purposefulness on the part of the assisting state. Neither was present here.[20]

Clearly, the defendant's justificatory strategies are interest-based. At the same time, they highlight the way in which the boundaries of what is deemed a responsible actor can shift in the presence of algorithmic infrastructures. In other words: where such infrastructures are involved, the classifiers and limits that define a responsible social actor become contested.

The Yemeni case was ultimately dismissed—although activist groups have recently indicated[21] that it may be referred to the Federal Court of Administration (Bundesverwaltungsgericht) for review. Despite the negative outcome, from the point of view of the purpose of 'strategic litigation'—namely, to create public awareness of clandestine state action, establish a public narrative about it, and generate relevant cultural resonance—the NGO strategy can be considered a success. By way of example: the court's ruling considers targeted killing by drone in the context of the violation of international law (though ultimately deciding, along with the defendant, that there is nothing to show that the U.S. military is guilty of this):

> It is true that drone operations sometimes involve the deaths of civilians, but this would only constitute a violation of international humanitarian law if the attacks were carried out indiscriminately or if, in the case of a strike against a legitimate military target, disproportionate levels of harm to the civilian population were factored in. In a speech delivered in May 2013, the President of the United States declared that drone operations would only take place if there was "near certainty" that no civilians would be killed in them. Such a standard would be compatible with international humanitarian law. The Federal Government works on the assumption that these guidelines are generally adhered to. It is inherently impossible, from a practical point of view, for [it] to check U.S. practice on drone warfare in every individual instance, given that the selection of drone targets takes places in strict secrecy.[22]

The success of strategic litigation emerges clearly here: the court makes reference to secret state practices, it sets these practices against the background of speeches made by the relevant head of state, and it considers in what way other states may be involved in a supportive capacity. It even discusses international humanitarian law and its prohibition on the indiscriminate and disproportionate use of force.

Scholars of philosophy and cultural sociology argue that narratives of justification are a prerequisite of any moral and just collective order (Forst 2014). Norms may appear abstract and fortuitous, but in fact they always evolve out of an ongoing narration of events in which protagonists are identified and good actions are differentiated from bad (Alexander 2006b). In the Ramstein case, the key element that made possible the strategic litigation was the presence of the algorithmic infrastructure. The narrative was not just about human actors, individual or collective; it also featured algorithms either as quasi-actors or as 'actants.' The latter is a term coined by literary theorists to denote objects that create or initiate a story. As the Ramstein case demonstrates, the role of such objects is not confined to the generation of literary narratives; they can also play a crucial part in legal argument.

Concluding remarks

The drone metaphor sits ill with classic actor theory and re-directs attention towards more hybrid entities such as 'actants.' Drones are actor-like—or actant-like—machines. They project the image of an autonomous actor—and simultaneously belie it. Seen as actants, drones create a fuzzily defined actor network that extends from one continent to another and enables states to act micro-globally.

But the drone metaphor has a cultural as well as a technical significance: it highlights the fact that the chain of transaction between people and action has become so great that the status of 'actor' can no longer be taken for granted. Drones have acquired their status of hybrid actor as a result of the proliferation in activating intermediaries called 'algorithms.' Algorithms thus mark the difference between those objects we perceive as vehicles and those we dub 'drones.'

The Euro Hawk project failed because the field of potential actors was 'zoomed in' to include only humans. The problem with this is that mobilizing objects act and interact in face-*off*-face situations while continuing to follow the logic of face-*to*-face situations. The Euro Hawk case demonstrates what might happen if action is still ultimately ascribed to humans seated at screens while algorithms—despite being an integral part of situations of interaction—are ignored. The case of Ramstein, conversely, involves a 'zoomed-out' concept of the actor: the legal proceedings were triggered not only by humans but by an algorithmic relay station and it was this which served as an entry point for civil rights activists to challenge clandestine drone operations. The strategic litigation succeeded in illuminating the existence of the algorithmic infrastructure.

Finally, both cases reveal the power of mobilizing algorithms to initiate narratives and cultural imaginings in regard to actors and the limits of acting entities. Social actors can no longer be assumed to take the form of contained, unvarying entities embodied only in the human frame. Their contours are becoming increasingly diffuse, expanding and contracting depending on the algorithmic interactions involved. That said, algorithmic agency should not be confused with the cultural imaginings born of human free will, intentionality, creativity, and freedom. Algorithmic autonomy is an autonomy that is embedded in interaction and infrastructures; it does not enable its possessor to invent something entirely new in a moment of transcendent creativity. In this sense, it is necessarily linked to other agents and is therefore only ever partial. Even so, as the two cases here demonstrate, mobilizing algorithms transgress the boundaries of what has traditionally been viewed as human action.

Notes

1 See e.g., www.transport.wa.gov.au/imarine/marine-radios.asp (accessed December 11, 2015).
2 Quotation from "Autonomous Weapons: An Open Letter from AI and Robotics Researchers," which opens as follows:

> Autonomous weapons select and engage targets without human intervention. They might include, for example, armed quadcopters that can search for and eliminate people meeting certain pre-defined criteria, but do not include cruise missiles or remotely piloted drones for which humans make all targeting decisions. Artificial Intelligence (AI) technology has reached a point where the deployment of such systems is—practically if not legally—feasible within years, not decades.

The letter, first published on July 28, 2015, is available online at http://futureoflife. org/open-letter-autonomous-weapons (accessed December 15, 2015).

3 C4ISR&networks, "After GPS: The Future of Navigation." March 31, 2015. www.c4isrnet.com/story/military-tech/geoint/2015/03/31/gps-future-navigation/70730572/ (accessed December 10, 2015).

4 This kind of system is mandatory for flight not only over the United States and Canada but also over EU territory: www.defensenews.com/story/defense/air-space/isr/2015/01/16/germany-euro-hawk-uas-/21799109/ (accessed January 16, 2015).

5 www.bundestag.de/dokumente/textarchiv/2013/46097693_kw30_ua_eurohawk_anhoerung/213238 (accessed November 24, 2015).

6 www.bundestag.de/dokumente/textarchiv/2013/45664684_kw26_ua_eurohawk/213064 (accessed November 24, 2015). All translations are by the author.

7 www.bundestag.de/dokumente/textarchiv/2013/46097693_kw30_ua_eurohawk_anhoerung/213238 (accessed November 24, 2015).

8 Attempts are currently being made to revive the German drone project. Whether the proposed new system, dubbed 'Triton,' will operate successfully has yet to be demonstrated, but critics have already expressed doubts on this score: www.merkur.de/politik/euro-hawk-ersatz-kostet-ueber-halbe-milliarde-euro-zr-5254055.html (accessed November 26, 2015).

9 www.reprieve.org.uk/ (accessed December 9, 2015).

10 www.ecchr.eu/de/home.html (accessed August 20, 2015).

11 Declaration [of] Faisal bin Ali Jaber, available at the website of the European Center for Constitutional and Human Rights: www.ecchr.eu/de/unsere-themen/voelkerstraftaten-und-rechtliche-verantwortung/drohnen/jemen.html (accessed September 1, 2015).

12 See www.generalbundesanwalt. de/docs/drohneneinsatz_vom_04oktober2010_mir_ali_pakistan (accessed January 10, 2016). Also www.ecchr.eu/de/unsere-themen/voelkerstraftaten-und-rechtliche-verantwortung/drohnen/pakistan.html (accessed January 10, 2016).

13 Verwaltungsgericht Köln, Urteil vom 27.05.2015, Aktenzeichen 3 K 5625/14. Abs. 4 (hereinafter cited only by paragraph). Full text at: www.justiz.nrw.de/nrwe/ovgs/vg_koeln/j2015/3_K_5625_14_Urteil_20150527.html (accessed February 10, 2016).

14 Abs. 4.

15 Abs. 4.

16 Abs. 9.

17 Abs. 11.

18 Abs. 16.

19 Abs. 16.

20 Abs. 16.

21 See www.ramstein-kampagne.eu/ (accessed February 2, 2016).

22 Abs. 87.

References

Alexander, J. C. 2006a. *The Meanings of Social Life: A Cultural Sociology*. New York: Oxford University Press.

Alexander, J. C. 2006b. "Cultural Pragmatics: Social Performance between Ritual and Strategy." In *Social Performance: Symbolic Action, Cultural Pragmatics and Ritual*. Edited by J. C. Alexander, B. Giesen, and J. Mast, 29–90. Cambridge, UK: Cambridge University Press.

Alexander, Jeffrey C. 2006c. *The Civil Sphere*. New York: Oxford University Press.

Anderson, C. W. 2013. "Towards a Sociology of Computational and Algorithmic Journalism." *New Media & Society* 15 (7): 1005–1021.

Arkin, R. C. 2007. "'Accountable Autonomous Agents': The Next Level." Position paper for the DARPA Complete Intelligence Workshop, February 2007.

Beer, D. 2009. "Power through the Algorithm? Participatory Web Cultures and the Technological Unconscious." *New Media & Society* 11 (6): 985–1002.

Bennett, W. L. and Segerberg, A. 2012. "The Logic of Connective Action." *Information, Communication & Society* 15 (5): 739–768.

Blum, G. and Heymann, P. 2010. "Law and Policy of 'Targeted Killing'." *National Security Journal* 1: 150–170.

Blumer, H. 1969. *Symbolic Interactionism: Perspective and Method.* Englewood Cliffs, NJ: Prentice-Hall.

Chamayou, G. 2015. *A Theory of the Drone.* New York: The New Press.

Daase, C. 2016. "On Paradox and Pathologies: A Cultural Approach to Security." In *Transformations of Security Studies.* Edited by G. Schlag, J. Junk, and C. Daase, 82–93. London: Routledge.

Daston, L. and Galison, P. 2007. *Objectivity.* New York and Cambridge, MA: MIT Press.

Ferran, Lee. 2014. "Ex-NSA Chief: 'We Kill People Based On Metadata'." *ABC News* May 12, 2014. Retrieved from http://abcnews.go.com/blogs/headlines/2014/05/ex-nsa-chief-we-kill-people-based-on-metadata/ (accessed October 11, 2015).

Forst, R. 2014. *The Right to Justification: Elements of a Constructivist Theory of Justice.* New York: Columbia University Press.

Fuchs, G. 2013. "Strategic Litigation for Gender Equality in the Workplace and Legal Opportunity Structures in Four European Countries." *Canadian Journal of Law & Society* 28 (2): 189–208.

Gerlitz, C. and Lury, C. 2014. "Social Media and Self-evaluating Assemblages: On Numbers, Orderings and Values." *Distinktion: Scandinavian Journal of Social Theory* 15 (2): 174–188.

Giesen, B. 2006. "Performing the Sacred: A Durkheimian Perspective on the Performative Turn in the Social Sciences." In *Social Performance: Symbolic Action Cultural Pragmatics and Ritual.* Edited by J. C. Alexander, B. Giesen, and J. Mast, 326–366. Cambridge, UK: Cambridge University Press.

Gillespie, T. 2010. "The Politics of 'Platforms'." *New Media & Society* 12 (3): 347–364.

Gillespie, T. 2014. "The Relevance of Algorithms." In *Media Technologies.* Edited by T. Gillespie, P. Boczkowski, and K. Foot, 167–194. Cambridge, MA: MIT Press.

Heintz, B. 1993. *Die Herrschaft der Regel. Zur Grundlagengeschichte des Computers.* Frankfurt/M and New York: Campus.

Höfera, T., Przyrembelb, H., and Verlegerc, S. 2004. "New Evidence for the Theory of the Stork." *Paediatric and Perinatal Epidemiology* 88: 88–92.

Hutchins, E. 1995. *Cognition in the Wild.* Cambridge, MA: MIT Press.

Introna, L. D. 2011. "The Enframing of Code: Agency, Originality and the Plagiarist." *Theory, Culture & Society* 28 (6): 113–141.

Introna, L. D. and Wood, D. 2004. "Picturing Algorithmic Surveillance: The Politics of Facial Recognition Systems." *Surveillance & Society* 3 (2): 177–198.

Junk, J. and Rauer, V. 2015. "Combining Methods: Connections and Zooms in Analyzing Hybrids." In *Transformations of Security Studies.* Edited by G. Schlag, J. Junk, and C. Daase, 216–232. London: Routledge.

Knorr-Cetina, K. 2005. "From Pipes to Scopes: 'The Flow Architecture of Financial Markets'." In *The Technological Economy.* Edited by Andrew Barry und Don Slater, 123–143. London: Routledge.

Knorr-Cetina, K. 2009. "The Synthetic Situation: Interactionism for a Global World." *Symbolic Interaction* 32 (1): 61–87.

Knorr-Cetina, K. and Bruegger, U. 2002. "Global Microstructures: The Virtual Societies of Financial Markets." *American Journal of Sociology* 107 (4): 905–950.

Knowles, C. and Burrows, R. 2014. "The Impact of Impact." *Etnográfica* 18 (2): 237–254.

Latane, B. and Darley, J. M. 1968. "Bystander Intervention in Emergencies: Diffusion of Responsibility." *Journal of Personality and Social Psychology* 8 (4): 377–383.

Latour, B. 2005. *Reassembling the Social: An Introduction to Actor-Network-Theory.* Oxford: Oxford University Press.

Latour, B. 2006. "Ethnologie einer Hochtechnologie." In *Technografie: Zur Mikrosoziologie der Technik.* Edited by W. Rammert and D. Schubert, 25–60. Frankfurt/M: Campus.

Lucas, G. R. 2013. "Engineering, Ethics, and Industry: The Moral Challenges of Lethal Autonomy." In *Killing by Remote Control.* Edited by B. J. Strawser, 211–228. Oxford: Oxford University Press.

Marres, N. 2012. *Material Participation: Technology, the Environment and Everyday Publics.* New York: Palgrave Macmillan.

Mast, J. 2011. "The Rise of Performance in Mass Democratic Politics." *Cahiers de recherche sociologique* 51: 157–180.

Matthias, A. 2004. "The Responsibility Gap: Ascribing Responsibility for the Actions of Learning Automata." *Ethics in Information Technology* 6 (3): 175–183.

Matthias, A. 2011. "Algorithmic Moral Control of War Robots: Philosophical Questions." *Law, Innovation and Technology* 3 (2): 279–301.

November, V., Camacho-Hu, E., and Latour, B. 2010. "Entering a Risky Territory: Space in the Age of Digital Navigation." *Environment and Planning D* 28 (4): 581–599.

Rauer, V. 2012a. "Interobjektivität. Sicherheitskultur aus Sicht der Akteur-Netzwerk-Theorien." In *Sicherheitskultur. Soziale und politische Praktiken der Gefahrenabwehr.* Edited by C. Daase, P. Offermann, and V. Rauer, 69–93. Frankfurt/M: Campus.

Rauer, V. 2012b. "The Visualization of Uncertainty." In *Iconic Power: Materiality and Meaning in Social Life.* Edited by J. C. Alexander, D. Bartmański, and B. Giesen, 139–154. New York: Palgrave Macmillan.

Sanz, E. and Stancik, J. 2014. "Your Search—'Ontological Security'—Matched 111,000 Documents: An Empirical Substantiation of the Cultural Dimension of Online Search." *New Media & Society* 16 (2): 252–270.

Schörnig, N. 2010. "Robot Warriors: Why the Western Investment into Military Robots Might Backfire." PRIF Report, no. 100. Frankfurt/M: Peace Research Institute Frankfurt/M.

Schulz-Schaeffer, I. 2007. *Zugeschriebene Handlungen. Ein Beitrag zur Theorie sozialen Handelns.* Weilerswist: Velbrück Wiss.

Schulz-Schaeffer, I. 2009. "Handlungszuschreibung und Situationsdefinition." *Kölner Zeitschrift für Soziologie und Sozialpsychologie* 61 (2): 159–182.

Simonson, S. 2011. *Rediscovering Mathematics.* Washington, D.C.: Mathematical Association of America.

Singer, P. W. 2015. "The Future of War Will Be Robotic." *CNN News*, February 23. Retrieved from http://edition.cnn.com/2015/02/23/opinion/singer-future-of-war-robotic/index.html (accessed May 20, 2016).

Sparrow, R. 2007. "Killer Robots." *Journal of Applied Philosophy* 24 (1): 62–77.

Sterio, M. 2012. "The United States' Use of Drones in the War on Terror: The (Il)legality of Targeted Killings under International Law." *Case Western Reserve Journal of International Law* 45 (1): 197–214.

Strawser, B. J., ed. 2013. *Killing by Remote Control*. Oxford: Oxford University Press.

Suchman, L. and Weber, J. Forthcoming 2016. "Human-Machine Autonomies." In *Autonomous Weapon Systems: Law, Ethics, Policy*. Edited by N. Buta, C. Kress, S. Beck, R. Geiss, and H. Y. Liu. Cambridge: Cambridge University Press.

Vöcking, B. 2011. *Algorithms Unplugged*. New York: Springer.

Vries, K. 2010. "Identity, Profiling Algorithms and a World of Ambient Intelligence." *Ethics and Information Technology* 12 (1): 71–85.

Williams, B. G. 2013. *Predators: The CIA's Drone War on Al Qaeda*. Dulles, VA: Potomac Books.

Wittgenstein, L. 1969. *On Certainty*. Oxford: Basil Blackwell.

9 Social bots as algorithmic pirates and messengers of techno-environmental agency

Oliver Leistert

To understand 'social bots' by affirming the 'social' in social bots uncritically incorporates and integrates a bundle of problems that stem from the name of the environment in which these bots nowadays operate: social media itself is a contested term originally launched with commercial rather than an uncoded, open sociality in mind. As such, social bots may be defined as "engag[ing] in two-way, direct communication with human users through natural language" (Graeff 2014, 2), often mimicking 'real' users (Hingston 2012). But 'real' and 'not-so real' is a rather naïve attempt at differentiation, as the 'real' users of commercial social media platforms ultimately are the customers of these platforms, for example advertisement brokers and surveillance agencies.

The abovementioned feature of natural language processing and production, even in its most primitive form, can be said to qualify for sociality in a basic sense, but such a starting point would inscribe bots into regimes of signification which have themselves been taken hostage by contemporary media technologies (Langlois *et al.* 2015).

This chapter therefore tries to understand bots more from the angle of their environment and as a part and parcel of such a media technological environment. At the same time, capital's imperative to feed on data extraction and to colonize the ephemeral uttering of 'people' also necessitate the situation of bots within these logics. As I will argue, bots on commercial platforms are (1) 'natural' inhabitants produced by the logics of the platforms and protocols themselves, and (2) a symptom of what might be called 'algorithmic alienation,' a process that currently redefines the very nature of knowledge production, as indicated throughout this whole book.

A relational existence in a digital milieu feeding on trust

Anthropocentric theories of the social may detest the idea that bots are social. Nonetheless, increasingly a general consensus has emerged stating that sociality is not an exclusive domain of the human (or animal being, for that matter) and that agency should be attributed to all kinds of beings and things, because it only becomes actual through a processual chain of nodes (Thrift 2005). Especially in media and technology studies, a shift can be noted from the question of 'what is' to the question of 'how' and 'who with whom,' indicating that instead of pinning down essentialist categories, a more nuanced focus on the questions of relations, operations and performativity is helpful.

In addition, the co-construction of technology and society has become a commonly followed trajectory (Boczkowski 1999). Recent anthologies of Science and Technology Studies that try to connect to media studies underline the necessity of overcome the idea of the 'what' in question (Gillespie *et al.* 2014), as has Actor-Network-Theory, which aims at symmetry in the description of technologies and societies, where mutual exchange instigates a plane of cross-polluting actants (cf. Latour 1999). This has led to a change in perspective and a transition from a fixation on static beings and objects towards relations and dynamics that are put to work by all kinds of agencies.

This comes in handy for the study of bots because what signifies bots without a doubt is their attempt to build relations, and not so much their ontological essence or static endurance. This is as true for the sophisticated social botnet on Facebook as it is for the rather simple spam bot. All bots aim at connections via data that either is initiated by humans or indexes humans, such as credit card information, or that is set up for humans, for example traps like phishing websites. Bots therefore are a mirror of our own captivity in machine-centered milieus, hinting at an amalgamation of technocultural socialities with networking infrastructures. They exemplify the necessity of understanding that trust on such platforms is of an algorithmic nature.[1] Trust then has become a relational property of computation, indexable and operational, in the end validated—or not—by humans. Trust has mutated into a discriminable parameter, proposed by machines. The algorithmic production of trust, with which bots engage and resonate, is built deeply into the platform's models of exploitation and wealth extraction. Trust in and through algorithmic powers delineates a new trajectory that has freed itself from the humanist concept of trust, and operates on premises of computability, the same milieu that social bots populate.

The commercial homes of social bots

To highlight the trajectory of current social bots, a sketch of the changes towards an algorithmic regime of media technologies under the engine of monetarization and capitalist enclosure is needed. The proliferation of social bots goes hand in hand with the proliferation of corporate social media platforms, which have dramatically changed and challenged the fabrics of sociality, including our understanding of the public (Baym and Boyd 2012), friendship (Bucher 2013), collective action (Dencik and Leistert 2015) and a shift in the datasphere in general (Langlois *et al.* 2015).

This development has been accompanied by a strong shift in privacy regulations and policy towards private actors as well (Braman 2006; Hintz 2015). The enmeshment of state and commercial surveillance (Landau 2010; Bauman and Lyon 2013) is one of the current key challenges in the relation between citizens and states (see www.dcssproject.net) posed by the unfettered rise of social media data empires. Ever since the Snowden revelations (Greenwald 2014), we now know how those platforms form an instrumental part of a surveillant assemblage (Haggerty and Ericson 2000) that has made operational as many heterogeneous sources of data as possible (Lyon 2014) for the production of new knowledge based on pattern recognition and correlation.

Among the many phenomena that emerged within these new algorithmic regimes is the struggle over collected data, and how and by whom data may be exploited. This remains an unresolved site of conflict on a truly global scale, as indicated by numerous juridical and political processes. One important aspect here relates to the ongoing privatization of communication, where everyday ephemeral utterings of millions of people becomes the retained property of the platform providers. This has been described as "digital enclosure" (Andrejevic 2007), that is, the enclosing of previously non-privatized communications for purposes of data mining and data selling. In addition, this enclosure plays back on the enclosed: it pre-formats, prescribes and designs the expressions on its platforms. Its powers are soft and the fact that censorship is part of the 'terms of service' signifies the hold that corporate platforms have on society and culture. It is this emergence of such database empires within the short span of a few years that naturally attracts players of all sorts who want their data share, beyond the official shareholders, agencies and advertisement brokers. Social bots are therefore just another kind of integrated player that queue up to gain access to the datafied fabric of social relations. This explains why a platform like Twitter has turned into a true botsphere where bots have become an integral element. In 2014, Twitter acknowledged that 8.5 percent of all its accounts are bots, or 23 million in absolute numbers, with an additional 5 percent of spam bots (Goldman 2014).

Beyond the scopes of legal or illegal: corporate platforms format the plane for bots

To shed some light on the emergence and role of social bots within Facebook's or Twitter's database empires, the legal discussions, including discussions of morality, that surround them, are not helpful, because under such terms bots always resonate within a predefined space, in the sense that they will always be portrayed as intruders to the regime that organizes this space. These networked software pieces are thus often described as troublemakers within the clean realms of enclosed and privatized database empires (Dunham and Melnick 2008), calling for legal regulation (Schellekens 2013) that aims at discriminating between regime bots and non-regime, malicious bots. Non-regime bots then are reminders that the promise made by companies such as Facebook to offer a safe and clean networked environment version of the otherwise 'dangerous' and open internet is impossible to accomplish as long as these remain radical neoliberal endeavors driven by capitalist interest. These platforms, in all their mutations and radicalizations of the networked logic, have brought the 'natural' predatory trajectories of capital's operations to a new schizophrenic intensity by their technologically driven integration of data accumulation and mining into an ambience of total surveillance. It is here that the production and operations of social bots have become an economy of their own, even an industry of their own, for example by providing anyone in need of attention and visibility with a high numbers of followers on Twitter in exchange for a few dollars (Messias *et al.* 2013). Bots feed on the virus that social media seeds into contemporary subjectivities: be visible, be ranked, be important, but only as an discriminable individual.

What is more, bots attack the currency of trust of such platforms. Not surprisingly, the discussion on social bots relates to issues of pollution of the public sphere by bots or questions of trust and believability (Hingston 2012), or points to the precarity of algorithmic measurements of users and their influence (Messias *et al.* 2013). The fact that research has investigated the users most likely susceptible to bots (Wagner and Strohmaier 2010) indicates another shift in networking responsibility within corporate realms, and reintroduces responsibility among users, for instance, by making users themselves responsible for connecting with bots, while ignoring that it is these very users, laboring unpaid under the 'protection' of the platform regimes, who are the source of monetary wealth for such platforms in the first place (Andrejevic 2011; Fuchs 2013).

It is this schizophrenic push by corporate platforms to appropriate the cognition and affects of billions, while producing capturing fields of desire, that lays the foundations for an arms race between bot developers (Boshmaf *et al.* 2011). As such, this instigates counter-measures, for instance, analyses that use the same algorithmic paradigms to determine if followers on Twitter are 'fake' or not (Bilton 2014), or, in the case of Facebook, the 'Facebook Immune System' (Stein *et al.* 2011), aiming at, among other things, containing bots and neutralizing them.

The problem of introducing a differentiation between good and bad bots is intrinsically linked to the problem of ownership of and access to data. Whether the bots are official ones or 'pirates,' they are excellent examples of a "fundamental uncertainty about who we are speaking to" (Gillespie 2014, 192) in times of algorithmically produced publics (Anderson 2012; Snake-Beings 2013).

The (reverse) Turing test can be applied to this uncertainty: in the fields of critical internet studies, social bots are seen as a mirror of our own reduction to machinelike actors within highly standardized environments. "Socialbots are a reflection of our activities within social media; for these machines to work, we ourselves have to be trained to be machinelike" (Gehl 2014, 16), which essentially means that we have become producers "of aggregated patterns of textually encoded, discrete states of mind" (Gehl 2014, 34). Discrete states (of mind) is the necessary precondition for computability and qualification for a Turing test. One has to conclude without any cynicism that the successful mobilization of large parts of populations to succumb to a Turing test essentially qualifies bots as equal partners.

The impossible catalogue of the botsphere

It remains challenging, maybe even impossible, to assign types to social bots, because it is a highly dynamic field that is strongly interdependent with the platforms and environments the bots run on. Nonetheless, I want to give a couple of examples and characterizations of bots, to show the importance and diversity of bots in today's internet assemblages.

Even the notion that bots run on platforms is misleading in many cases, because bots may well run on decentralized servers in connection with platforms, as is the case with Wikipedia-affiliated bots (Geiger 2014); this points to the necessity of focusing on relational epistemologies instead of types. In addition, since many bots are considered rogue or malicious, the means for identifying them, of shutting them down or containing them in other ways, like channeling them into pure 'bot land' or programming anti-bot bots, have generated their own research fields, triggering intense competition between bot programmers and counter-bot programmers (Wang 2010). Such competitions dramatically introduce complexity into bot milieus; it becomes more precarious than ever to regard bots as objects with clear boundaries. Bots can be highly flexible, changing their behavior, and even (machine-)learning (Boshmaf *et al.* 2012), leading to further adaptation techniques. Their milieus add to this complication with their standardized input forms, handling of strings, for instance in sentiment analysis, and databased calculations of relations. The more the internet becomes a culture of templates and standardized web interfaces, the easier it is to simulate agility and vitality, since these are already curbed and mutated to be processable and mine able. With reference to Baudrillard, one might say that the simulation bears its own (bot) children.

The technical description of bots as semi- or fully automated agents does not say much about the role they play in datafied capitalist environments either. Bots are so much more than their code (as is true of all software).Their elegance and agency only becomes significant when their code is executed in a networked environment. I agree with Geiger that "bots are a vivid reminder that what software is as software cannot be reduced to code and divorced from the conditions under which it is developed and deployed" (Geiger 2014, 346). As I will suggest later, the figure of the algorithmic pirate for social bots might be a more satisfying approach since it allows us to situate bots within the field of a political economy, a perspective that adds another layer onto the milieu of bots while at the same time providing a metaphorical and thus *signifying description* for otherwise asignifying machines.

I propose two criteria, *purpose* and *software*, as means of differentiation within this dynamic field of the botsphere. 'Purpose' asks what the bots programmed goal is, and suggests an examination of their performance in and relation within and to their milieu. 'Software' is shorthand for their technical implementation, which, on a more detailed description and interpretation, would necessarily include libraries, technological standards, networking capacities, programming languages and the machines that host them, including the hardware.

Beyond these components, a description of a bot assemblage aiming at integrity must include the production and coding process, such as programmer's exchanges and communications, and iterations and adjustments during operations, for example by the herder of a botnet. In addition, the proclaimed automaticity of bots needs critical investigation too, because often an update is manually made in response to changes in the bot's milieu.

Examples of bots, more or less social

The first of these semi- to fully automated networked software pieces then is *chatter bots* that are used for purposes that are seen as being too tedious for contemporary humans to perform, but that nonetheless remain necessities that need to be performed in their respective fields. Such are Internet Relay Chat (IRC) bots, who "sit on specific channels, to enforce channel rules and policies by monitoring public conversations, and to take actions against those violating channel rules, as well as to give certain individuals operator status on request" (Latzko-Toth 2014, 588). IRC bots, as Latzko-Toth explains, are programmed on top of the IRC software, which is their milieu, as ancillary code, and perform management and maintenance tasks, thus providing help in governance of the channels. The level of sophistication can vary and is finally determined by the IRC software itself, which then can lead to an uncontrolled growth of patches for this open source software to offer new functionalities for bots.

Like IRC bots, bots on Wikipedia are also little helpers which perform essentially two tasks. First, like IRC bots, they algorithmically help govern the large and international Wikipedia community. I propose we call them *governor bots*. Their purpose is to "make it possible to achieve a certain level of uniformity in style and content," and "also serve key governance roles, independently enforcing discursive and epistemological norms, particularly for newcomers" (Geiger 2014, 345). For the English Wikipedia, there are, at the time of writing, 1903 "bot tasks approved for use" (Wikipedia 2015a). A bot approval group watches over this armada of semi- to fully automatized agents.

Second, bots on Wikipedia are actually 'writing' articles. For example, "[t]he so-called *rambot* operated by Ram-Man created approximately 30,000 U.S. city articles based on U.S. Census tables, at a rate of thousands of articles per day" (Wikipedia 2015b). The content providing bots' activities vary very much across each language's Wikipedia incarnation.[2]

A relative of the Wikipedia content producing bot is the *review 'writing' bot* on shopping portals or recommendation sites. Since these programs intend to spur the sales of a specific product by simulating a real experience with the product, they share some features with the figure of the pirate as they try to manipulate opinions on products and ultimately money flows. But these features, one may argue, are internal to market logics anyway.

The *recommendation bot* thus can be understood as rogue or benevolent, depending on its milieu. On Amazon, most users would see such a bot as helpful or, at least, would ignore it, while on the many price comparison sites, such bots are often considered as rogue, or at least illegitimate. Still, its customers who have paid for its deployment are often linked directly or indirectly to sales and promotion agencies or the producers themselves. Recommendation bots are natural occurrences in a data-driven economy. Their purpose is to facilitate the link between commodity data and money data. The connection between these two data categories that they try to build is the *ultimate connection*.

The next candidate in this brief overview of bots also operates in the realm of opinion production and manipulation, but with different intent. *Sock puppet bots* can emerge in different milieus, from the Twitter sphere to Reddit. Their purpose is to influence ongoing debates, which may even involve the destruction of communication spheres. The term usually used to refer to these propaganda operations is astroturfing (Leistert 2013). The means they have at their disposal vary, from flooding lively debates with recurrent messages, which simply disrupts the possibility of continued discussions and exchange of opinions, to the sophisticated targeting of identified opinion leaders and vilifying them. Such bots provide an alternative to censorship as they induce huge amounts of noise and make for instance political discussions impossible while the communication channels as such are not switched off. Government agencies, religious entities and corporations alike operate such bots—not only in times of official crisis and conflict. In sum, sock puppet bots induce disruptive vectors into algorithmically produced public spheres, which relates them to a branch of warfare called PsyOps (Paganini 2013).

Well known and most prominently operating on Twitter are bots that follow users to enhance the user's fame and popularity. Although this is a risky strategy since observers often identify a sudden rise in popularity, it is applied broadly across very different societal strata. Politicians and pop stars are common customers of these *fame enhancing bots*. But such bots have also become a common means in marketing and public relations to spur the popularity of brands or products. *Fame enhancing bots* take advantage of the platform's logics of capture, operating on the contemporary societal imperative that everyone's opinion and sentiment counts, leading to the incorporation of dividing trajectories among users: to be is to be visible (Milan 2015), hence to be is to be *more* visible than all the others. Such bots take care of this 'more,' because they are designed to replicate fast.

Harvesters are among those bots that I propose to call *algorithmic pirates* in the fullest sense of the term. They infiltrate social networking sites and try to friend as many users as possible on Facebook. They harvest as much personal data as possible from them while remaining undiscovered as bots. Set up with sophisticated profiles and activity patterns that resemble humans, they attack the business models of social media platforms at their core since they leak to their herder what is officially available for sale from the platform providers: data by and about users of the platform. Such bots are disguised. Their existence hinges on their camouflage, because they engage with humans as humans, whereas *fame enhancing bots* only add themselves to a list (of followers).

Clearly programmed with bad intentions by their cultivators are bots that try to insert malicious code into users' applications, such as redirecting users to phishing websites. The milieu of these bots is flexible. These *malicious bots* can prey on dating platforms or on Twitter alike. Their activity is punctual: once the target has followed the inserted link, their purpose is fulfilled. Nonetheless, to meet this goal, they often have to bond with users to achieve credibility. Such bots are algorithmic pirates too, because they use trusted environments, such as dating platforms, to divert traffic into prearranged traps for economic purposes.

Social bots as algorithmic pirates of data capitalism

Seen from the perspective of political economy, social bots can be described, with some modifications, as refurbished incarnations of the figure of the pirate, belonging to what Lawrence Liang calls "an entire realm that is inhabited by figures such as the trickster, the copier, the thief" (Liang 2010, 361). Modeling social bots as algorithmic pirates, which, as the metaphor suggests, are immersed in a sea of social media data, allows for a change in perspective on the database empires of commercial social media enterprises, because it decenters and thus re-questions the normalized perception of ownership of data. This includes a new angle of problematization on the data processing platforms themselves. Social bots, rogue or not, provide means to re-evaluate main strands of our contemporary data economy since they fragment and de/revalorize both the invested unpaid affective labor of contemporary post-fordist subjectivity (Ross 2013) and the contemporary business model of wealth extraction which these corporate social media platforms execute, by rechanneling labor and wealth into a "marginal site of production and circulation" (Liang 2010, 361). My argument is that social bots remodel/rechannel the established circuits into the cycles of unpaid labor and wealth extraction by running on or connecting to corporate infrastructures: they emerge as the suppressed 'other,' as what has been cleaned out in centrally governed walled realms, but that continues to haunt these realms—a doppelgänger of a very specific kind.

Before further discussing the operational trajectory of social bots from such a perspective, a brief reiteration of the discourse on media privacy is necessary in order to frame the explanatory limits of the figure of the algorithmic pirate.

Media technologies and the figure of the pirate

Media piracy is here to stay, and it is neither new nor an anomaly of capitalist ventures. It is only via the transition from Gutenberg's to Turing's media regime that media piracy has become such a scrutinized and contested topic within the last decades. Everyday mass scale, semi- and fully automated copying and redistribution of software, books, music and films on- and offline has put enormous pressure on a regime of copyright that originates from analogue times and from a different historical phase of capitalist expansion. Its consolidation and institutionalization, in its most emblematic and pathetic formation as the World Intellectual Property Organization (WIPO) in 1970, runs parallel to two historical events: the final end of official colonialism as a political regime, leading to new nation states that pursue their own agenda, and the connected unification of markets into an institutionalized 'world market,' led by the U.S. and Western Europe. Hegemony over intellectual property regimes since then has become a cornerstone of the powers of 'developed' countries over others.

This was different in the nineteenth century when the U.S. still had to catch up with European industrial development. To do so, it has been a common practice for American industry to copy European patents and other intellectual goods, and thus infringe on their 'intellectual property,' in order to compete with the developed European industry (Ben-Atar 2004). Today, it is most prominently the Americans themselves who try to guard against countries whose industries try to catch up by using rather rogue practices of copy and paste, like China or India. This shows how *natural* the piracy process should be understood historically within capitalist developments, as any "analyses of piracy delineate the boundaries and (il)legitimacies of specific regimes of power" (Zehle and Rossiter 2014, 345).

The urgency of finding solutions to the restrictions on access to information and immaterial goods is expressed through many lively debates on the commons (Linebaugh 2013) and the growing numbers of alternative copyright regimes, such as Creative Commons (Lessig 2002), or viral software licenses, such as the GNU license family. The ongoing efforts and initiatives to find a balance between a right to information and to immaterial goods and the interests of the not quite almighty copyright infringement persecutors, such as the RIAA,[3] and the many failed attempts to find a technological fix for a problem of what is ultimately an issue of social justice, such as Digital Rights Management, are a reminder how digital cultures operate under entirely different conditions as compared to previous regimes. If "power through the algorithm is increasingly important for media companies in digital rights management" (Lash 2007, 71), agency has without a doubt shifted in the digital milieu towards a new diversity of actors and actants.

To enforce intellectual property, a powerful and frightening discourse on piracy has been established where pirates are "seen as the ultimate embodiment of evil. That evil takes a variety of forms, from terrorism and the criminal underworld to causing the decline of the entertainment industry and evading of taxes" (Liang 2010, 356). Interestingly, at the same time, for actors "who work on limiting the expansion of intellectual property rights and on defending the public domain, the figure of the pirate is treated with embarrassed silence or outright disavowal" (Liang 2010, 356) Piracy, it seems, is driven by logics that transgress, trickster-like, the well-ordered Western concepts of ownership and legality. Thus, pirates will never be truthful allies of the commoners either. What is more, piracy is seen as a danger to the creative class, ultimately as an enemy of the artist, whose legitimate juridical claims on her inventions are ignored by pirates. Piracy transgresses the order of the 'knowledge economy' by ignoring its reterritorializing forces. This allegedly parasitic character of piracy misses *piracy's productive side*: piracy has invented across separated markets different highly creative forms of distribution (Maigret and Roszkowska 2015). This is the striking parallel to the way social bots redistribute data. The difference, though, is the mode of redistribution. Data caught by bots may go in very different directions, from credit card fraud to hacking websites, from strikingly fitting advertisements or spam to identity theft. Still, all the modes of redistribution and new ways to make the captured data operational again are potentially open for interested parties and buyers of any kind, exactly like so-called 'pirated media.'

Pirate bots, algorithms and infrastructure

The power of the pirate bots stems from the powers of the infrastructures they inhabit. It is the dissemination of powers into infrastructures, where modulations of control, famously sketched by Deleuze in his "Postscript on the Societies of Control" (Deleuze 1992), materialize as dispersed and decentralized logistical nodes, providing the proliferation of pockets of environmental agency, from which pirate bots profit, because "[p]iracy suggested not just a permanent loss of space and markets, but also a model of dispersal where 'distribution' took on a productive form. As distributor pirates also produced more media, piracy bred further piracy" (Sundaram 2009, 116) Once algorithms have become governors (Ziewitz 2015), functioning as analytics (Amoore and Piotukh 2015), pirate bots become nodes or elements of an assemblage that modulates infrastructural operations. Piracy nests in what Keller Easterling calls "extrastatecraft" (Easterling 2014) to describe the physical and non-physical formatting powers of the current political infrastructural economy, which "signals the imperial ambition of both standards and infrastructure" (Zehle and Rossiter 2014, 349). This is the milieu in which

> ... [p]iracy exists in commodified circuits of exchange, only here the same disperses into the many. Dispersal into viral swarms is the basis of pirate proliferation, disappearance into the hidden abodes of circulation is the secret of its success and the distribution of profits in various points of the network,...
>
> (Sundaram 2009, 137)

just as pirate bots are proliferating by swarming and later disappear.

Hence, piracy should be understood as complementary rather than "simply parasitical" to a "mode of relation that underwrites the resilience (and redundancy) of network infrastructures" (Zehle and Rossiter 2014, 348). For instance, for companies like Microsoft, software piracy has been a powerful strategy against open source software alternatives.

Beyond moral or legal considerations social bots can be situated in a data driven capitalist economy that itself threatens privacy and self-determination. Today's internet economy is by and large run by data brokers and advertising agencies, having successfully implemented the 'free as in free beer' logic, where monetization is based on opaque data trade that users know nothing about. Social bots are the suppressed complementary side to this business model. While media piracy has become a combat on many levels, data piracy is the next stage of an economy driven by data. Successfully attacking the currency of trust in social networking sites, bots also make visible the high degree of *algorithmic alienation* that social media platforms produce. The fact that these agents can successfully present themselves as human users signifies the precarious state of sociality on which social media hinges.

It is telling that these dark sides of internet culture are currently neglected in research. Beyond the few great exceptions (Parikka 2007; Parikka and Sampson 2009; Brunton 2013), it seems that research itself is blinded by the tales of efficiency and immateriality of net cultures, adding to the idealist discourse about the net. A materialist approach would have to start with neglected figures such as social bots as data pirates or, to give another all too often forgotten example, the fact that the porn industry has become a key player in the development of standards, for example in streaming technologies and the routing of traffic (Chun 2005). The internet as a gigantic machine to "pump up noise levels" (Lovink 2005, 10) has many actants, social bots among them.

In addition, moral or legal discussion cannot provide insights into the material nature of the internet. Facebook's ban on breastfeeding mothers, or their censorship system in general, though legal as such because it largely accords with their terms of service, is a good case in point: it demonstrates that algorithms have become co-governors of signifying expressions. They enforce what is possible and what is not in the net, and they also prescribe effectively the meaning of posts in their very limited and normative concepts of words—the poet's nightmare, so to speak.

Social bots then are variants of such algorithmic governors, but are unofficial or rogue ones, equipped with less systemic agency than their fellow platform algorithms, following their own programmed rules, ignoring the rules of the platforms they inhibit. To see them as parasites only comprehends half of this phenomenon. They are the complement of the current data economy, made possible by current business models themselves. The trust they are said to betray is itself disputable, because the computation processes that set it up are opaque to the users too. The delegation of such elementary desires of human beings to machine intelligence inevitably produces repercussions in forms such as social bots from the machinic logic itself. They only follow the trend that corporate platforms themselves have brought to a new level with their capture of the private/public distinction. Finally, "[a]lgorithmic technologies […] tend to reduce differences in kind to differences in degree […] or to distances data-point to data-point" (Amoore and Piotukh 2015, 361). Such a flattened sociality that is differentiated by degree only inevitably produces more of the same because of its fabulous unlimited capacities of reproduction and transmissions. Dynamically swapping its simulated points of reference, reappearing hydra-like after containment with new automatically produced profiles, social bots are the true first 'native inhabitants' of current social media environments, reproducing themselves generically following their preprogrammed reproduction rules.

To ban and contain them will remain unfeasible in the long run, because the platforms themselves have produced human users that provide the compatibility needed for bots to resonate with them. Being governed by the platform's standardizations, censor systems, terms of service, templates, algorithmic processes and databased connections is very much the logic of control through protocols (Galloway 2004). Algorithmic pirates, such as social bots, easily serve the protocol's needs and thus naturally are becoming active parts of such a techno-environmental logic of control.

Notes

1 This is not to be confused with trust based on cryptography, which responds to a related problem—trust in networked environments—but with means that are in the hands of the trusting parties involved.
2 Wikipedia and bots form a complex ecosystem, for further reading, see e.g., Geiger (2014) and Niederer and van Dijck (2010).
3 The Recording Industry Association of America (RIAA) is one of many organizations that watches over and enforces with disputatious means copyright infringements, among other things.

References

Amoore, L. and Piotukh, V. 2015. "Life beyond Big Data: Governing with Little Analytics." *Economy and Society* 44 (3): 341–366. doi: 10.1080/03085147.2015.1043793.

Anderson, C. W. 2012. "Towards a Sociology of Computational and Algorithmic Journalism." *New Media & Society* 15 (7): 1005–1021. doi: 10.1177/1461444812465137.

Andrejevic, M. 2007. "Surveillance in the Digital Enclosure." *The Communication Review* 10 (4): 295–317. doi: 10.1080/10714420701715365.

Andrejevic, M. 2011. "The Work That Affective Economics Does." *Cultural Studies* 25 (4–5): 604–620. doi: 10.1080/09502386.2011.600551.

Bauman, Z. and Lyon, D. 2013. *Liquid Surveillance a Conversation*. Cambridge, UK and Malden, MA: Polity Press.

Baym, N. K. and Boyd, D. 2012. "Socially Mediated Publicness: An Introduction." *Journal of Broadcasting & Electronic Media* 56 (3): 320–329. doi: 10.1080/0883 8151.2012.705200.

Ben-Atar, D. S. 2004. *Trade Secrets: Intellectual Piracy and the Origins of American Industrial Power*. New Haven, CT: Yale University Press.

Bilton, N. 2014. "Social Media Bots Offer Phony Friends and Real Profit." *New York Times*, November 19, 2014. Retrieved from www.nytimes.com/2014/11/20/fashion/social-media-bots-offer-phony-friends-and-real-profit.html?_r=0 (accessed October 24, 2015).

Boczkowski, P. 1999. "Mutual Shaping of Users and Technologies in a National Virtual Community." *Journal of Communication* 49 (2): 86–108. doi: 10.1111/j.1460-2466. 1999.tb02795.x.

Boshmaf, Y., Muslukhov, I., Beznosov, K., and Ripeanu, M. 2011. "The Socialbot Network: When Bots Socialize for Fame and Money." In *Proceedings of the 27th Annual Computer Security Applications Conference*, 2011, ACM, 93–102. Retrieved from http://dl.acm.org/citation.cfm?id=2076746 (accessed February 16, 2015).

Boshmaf, Y., Muslukhov, I., Beznosov, K., and Ripeanu, M. 2012. "Key Challenges in Defending against Malicious Socialbots." In *Proceedings of the 5th USENIX Conference on Large-Scale Exploits and Emergent Threats*, 2012, USENIX Association, 12–12. Retrieved from: www.usenix.org/system/files/conference/leet12/leet12-final10. pdf (accessed September 24, 2015).

Braman, S. 2006. *Change of State: Information, Policy, and Power*. Cambridge, MA: The MIT Press.

Brunton, F. 2013. *Spam: A Shadow History of the Internet*. Cambridge, MA: The MIT Press.

Bucher, T. 2013. "The Friendship Assemblage: Investigating Programmed Sociality on Facebook." *Television & New Media* 14 (6): 479–493.

Chun, W. H. K. 2005. *Control and Freedom: Power and Paranoia in the Age of Fiber Optics*. Cambridge, MA: The MIT Press.

Deleuze, G. 1992. "Postscript on the Societies of Control." *October* 59: 3–7.

Dencik, L. and Leistert, O., eds. 2015. *Critical Perspectives on Social Media and Protest: Between Control and Emancipation*. London: Rowman & Littlefield.

Dunham, K. and Melnick, J. 2008. *Malicious Bots: An Inside Look into the Cybercriminal Underground of the Internet*. Boca Raton, FL, London and New York: CRC Press.

Easterling, K. 2014. *Extrastatecraft: The Power of Infrastructure Space*. London and New York: Verso.

Fuchs, C. 2013. "Class and Exploitation in the Internet." In *Digital Labor: The Internet as Playground and Factory*. Edited by Trebor Scholz, 211–223. New York: Routledge.

Galloway, A. 2004. *Protocol: How Control Exists after Decentralization*. Cambridge, MA: The MIT Press.

Gehl, R. W. 2014. *Reverse Engineering Social Media: Software, Culture, and Political Economy in New Media Capitalism*. Philadelphia, PA: Temple University Press.

Geiger, R. S. 2014. "Bots, Bespoke, Code and the Materiality of Software Platforms." *Information, Communication & Society* 17 (3): 342–356. doi: 10.1080/1369118X. 2013.873069.

Gillespie, T. 2014. "The Relevance of Algorithms." In *Media Technologies: Essays on Communication, Materiality, and Society*. Edited by T. Gillespie, P. J. Boczkowski, and K. A. Foot, 167–193. Cambridge, MA: The MIT Press.

Gillespie, T., Boczkowski, P. J., and Foot, K. A., eds. 2014. *Media Technologies: Essays on Communication, Materiality, and Society*. Cambridge, MA: The MIT Press.

Goldman, D. 2014. "23 Million Twitter Users Are Fed by Robots." *CNN Money*. Retrieved from http://money.cnn.com/2014/08/12/technology/social/twitter-bots/index. html (accessed November 1, 2015).

Graeff, E. C. 2014. "What We Should Do Before the Social Bots Take Over: Online Privacy Protection and the Political Economy of Our Near Future." Massachusetts Institute of Technology, Cambridge, MA. Retrieved from http://web.mit.edu/sts/Graeff. pdf (accessed October 7, 2015).

Greenwald, G. 2014. *No Place to Hide: Edward Snowden, the NSA and the Surveillance State*. London: Penguin Books.

Haggerty, K. D. and Ericson, R. 2000. "The Surveillant Assemblage." *British Journal of Sociology*, 51 (4): 605–622. doi: 10.1080/00071310020015280.

Hingston, P., ed. 2012. *Believable Bots*. Berlin and Heidelberg: Springer.

Hintz, A. 2015. "Social Media Censorship, Privatized Regulation and New Restrictions to Protest and Dissent." In *Critical Perspectives on Social Media and Protest: Between*

Control and Emancipation. Edited by L. Dencik and O. Leistert, 109–126. New York: Rowman & Littlefield.

Landau, S. 2010. *Surveillance or Security?: The Risks Posed by New Wiretapping Technologies.* Cambridge, MA: The MIT Press.

Langlois, G., Redden, J., and Elmer, G., eds. 2015. *Compromised Data: From Social Media to Big Data.* New York and London: Bloomsbury.

Lash, S. 2007. "Power after Hegemony: Cultural Studies in Mutation?" *Theory, Culture & Society*, 24 (3): 55–78. doi: 10.1177/0263276407075956.

Latour, B. 1999. *Pandora's Hope: Essays on the Reality of Science Studies.* Cambridge, MA: Harvard University Press.

Latzko-Toth, G. 2014. "Users as Co-Designers of Software-Based Media: The Co-Construction of Internet Relay Chat." *Canadian Journal of Communication* 39 (4): 577–595.

Leistert, O. 2013. "Smell the Fish: Digital Disneyland and the Right to Oblivion." *First Monday*, 18 (3). doi: 10.5210/fm.v18i3.4619.

Lessig, L. 2002. *The Future of Ideas: The Fate of the Commons in a Connected World.* New York: Vintage.

Liang, L. 2010. "Beyond Representation: The Figure of the Pirate." In *Access to Knowledge in the Age of Intellectual Property.* Edited by A. Kapczynski and G. Krikorian, 353–376. New York: Zone Books.

Linebaugh, P. 2013. *Stop: The Commons, Enclosures, and Resistance.* Oakland, CA: PM Press.

Lovink, G. 2005. *The Principle of Notworking: Concepts in Critical Internet Culture.* Amsterdam: Amsterdam University Press.

Lyon, D. 2014. "Surveillance, Snowden, and Big Data: Capacities, Consequences, Critique." *Big Data & Society* 1 (2). doi: 10.1177/2053951714541861.

Maigret, N. and Roszkowska, M., eds. 2015. *The Pirate Book.* Ljubljana: Aksioma – Institute for Contemporary Art. Available at: http://thepiratebook.net.

Messias, J., Schmidt, L., Oliveira, R., and Benevenuto, F. 2013. "You Followed My Bot! Transforming Robots into Influential Users in Twitter." *First Monday*, 18 (7). doi: 10.5210/fm.v18i7.4217.

Milan, S. 2015. "Mobilizing in Times of Social Media: From a Politics of Identity to a Politics of Visibility." In *Critical Perspectives on Social Media and Protest: Between Control and Emancipation.* Edited by L. Dencik and O. Leistert, 53–70. London: Rowman & Littlefield.

Niederer, S. and van Dijck, J. 2010. "Wisdom of the Crowd or Technicity of Content? Wikipedia as a Sociotechnical System." *New Media & Society* 12 (8): 1368–1387. doi: 10.1177/1461444810365297.

Paganini, P. 2013. *PsyOps and Socialbots: InfoSec Resources.* Retrieved from http://resources.infosecinstitute.com/psyops-and-socialbots/ (accessed October 29, 2015).

Parikka, J. 2007. *Digital Contagions: A Media Archaeology of Computer Viruses.* New York: Peter Lang.

Parikka, J. and Sampson, T. D. 2009. *The Spam Book: On Viruses, Porn, and Other Anomalies from the Dark Side of Digital Culture.* New York: Hampton Press.

Ross, A. 2013. "In Search of the Lost Paycheck." In *Digital Labor: The Internet as Playground and Factory.* Edited by T. Scholz, 13–32. New York: Routledge.

Schellekens, M. H. M. 2013. "Are Internet Robots Adequately Regulated?" *Computer Law & Security Review* 29 (6): 666–675. doi: 10.1016/j.clsr.2013.09.003.

Snake-Beings, E. 2013. "From Ideology to Algorithm: The Opaque Politics of the Internet." *Transformations: Journal of Media and Culture* 23. Retrieved from www.transformationsjournal.org/issues/23/article_03.shtml (accessed May 28, 2016).

Stein, T., Chen, E., and Mangla, K. 2011. "Facebook Immune System." In *SNS 2011 Proceedings of the 4th Workshop on Social Network Systems*, ACM, 1–8. doi: 10.1145/1989656.1989664.

Sundaram, R. 2009. *Pirate Modernity: Delhi's Media Urbanism*. London and New York: Routledge.

Thrift, N. J. 2005. *Knowing Capitalism*. London: Sage.

Wagner, C. and Strohmaier, M. 2010. "The Wisdom in Tweetonomies: Acquiring Latent Conceptual Structures from Social Awareness Streams." In: *SEMSEARCH '10 Proceedings of the 3rd International Semantic Search Workshop*, 2010, ACM, 6. Retrieved from http://dl.acm.org/citation.cfm?id=1863885 (accessed November 28, 2014).

Wang, A. H. 2010. "Detecting Spam Bots in Online Social Networking Sites: A Machine Learning Approach." In *Data and Applications Security and Privacy XXIV*, Springer, 335–342. Retrieved from http://link.springer.com/chapter/10.1007/978-3-642-13739-6_25 (accessed November 28, 2014).

Wikipedia. 2015a. "Bots." https://en.wikipedia.org/wiki/Wikipedia:Bots (accessed October 25, 2015).

Wikipedia. 2015b. "History of Wikipedia Bots." https://en.wikipedia.org/wiki/Wikipedia:History_of_Wikipedia_bots (accessed October 25, 2015).

Zehle, S. and Rossiter, N. 2014. "Privacy Is Theft: On Anonymous Experiences, Infrastructural Politics and Accidental Encounters." In *Piracy: Leakages from Modernity*. Edited by J. Arvanitakis and M. Fredriksson, 345–353. Sacramento, CA: Litwin Books.

Ziewitz, M. 2015. "Governing Algorithms: Myth, Mess, and Methods." *Science, Technology & Human Values*. Published online before print September 30, 2015, doi: 10.1177/0162243915608948.

Index

Page numbers in **bold** denote figures.